I0482778

Standard Review Plan for the Review of a Reclamation Plan for Mill Tailings Sites Under Title II of the Uranium Mill Tailings Radiation Control Act of 1978

Final Report

U.S. Nuclear Regulatory Commission
Office of Nuclear Material Safety and Safeguards
Washington, DC 20555-0001

AVAILABILITY OF REFERENCE MATERIALS
IN NRC PUBLICATIONS

NUREG-1620, Rev. 1

Standard Review Plan for the Review of a Reclamation Plan for Mill Tailings Sites Under Title II of the Uranium Mill Tailings Radiation Control Act of 1978

Final Report

Date Completed: June 2003
Date Published: June 2003

Prepared by
J. Lusher

J. Lusher, NRC Project Manager

Division of Fuel Cycle Safety and Safeguards
Office of Nuclear Material Safety and Safeguards
U.S. Nuclear Regulatory Commission
Washington, DC 20555-0001

ABSTRACT

A U.S. Nuclear Regulatory Commission source and byproduct materials license is required by 10 CFR Part 40 for the operation of uranium mills and the disposal of "tailings," wastes produced by the extraction or concentration of source material from ores processed primarily for their source material content. Appendix A to Part 40 establishes technical and other criteria relating to siting, operation, decontamination, decommissioning, and reclamation of mills and of tailings at mill sites. The licensee's site reclamation plan documents how the proposed activities demonstrate compliance with the criteria in Appendix A to Part 40 and the information needed to prepare the environmental assessment on the effects of the proposed reclamation activities on the health and safety of the public and on the environment.

This standard review plan is prepared for the guidance of staff reviewers in the Office of Nuclear Material Safety and Safeguards in performing safety and environmental reviews of reclamation plans for uranium mill tailings sites covered by Title II of the Uranium Mill Tailings Radiation Control Act of 1978 as amended. It provides guidance for new reclamation plans, license renewals, and license amendments. The principal purpose of this standard review plan is to ensure the quality and uniformity of staff reviews and to present a well-defined base from which to evaluate changes in the scope and requirements of a review.

This standard review plan is written to cover a variety of site conditions and reclamation plans. Each section contains a description of the areas of review, review procedures, acceptance criteria, and evaluation findings. Revision 1 also incorporates information to address new Commission policy on several issues related to uranium recovery.

Paperwork Reduction Act Statement

Public Protection Notification

CONTENTS

CONTENTS (continued)

CONTENTS (continued)

CONTENTS (continued)

CONTENTS (continued)

CONTENTS (continued)

FIGURE

TABLES

EXECUTIVE SUMMARY

A U.S. Nuclear Regulatory Commission (NRC) source and byproduct materials license is required in accordance with the provisions of Title 10 of the U.S. Code of Federal Regulations, Part 40 (10 CFR Part 40), "Domestic Licensing of Source Material," in conjunction with uranium or thorium milling, or with byproduct material at sites formerly associated with such milling. At the termination of a uranium mill license, the mill tailings impoundment and some adjoining land will be turned over to the U.S. Department of Energy (DOE), another Federal agency designated by the President, or the State in which the site is located for long-term care. Requirements applicable to a license consist of the regulations in 10 CFR Part 40, Appendix A to 10 CFR Part 40, and any license conditions. The specific sections in this standard review plan that address the criteria of 10 CFR Part 40, Appendix A, are shown in Appendix A of the review plan.

An application for a new license, license renewal, or an amendment to or termination of an existing license should contain, as appropriate, proposed specifications relating to the milling operations, and the information on the disposal of tailings or wastes resulting from such milling activities and information on decommissioning of the site. General guidance on (i) contents and filing of an application and (ii) producing an environmental report appears in 10 CFR 40.31, "Application for specific licenses," and in 10 CFR 51.45, "Environmental report," respectively. The staff uses the information in the application to determine whether the proposed activities will be protective of public health and safety and be environmentally acceptable. General provisions for issuance, amendment, transfer, and renewal of licenses are described in 10 CFR Part 2, Subpart A. Guidance on considering environmental justice issues during licensing of Title II uranium or thorium mills is presented in NUREG–1748 (NRC, 2001).

This standard review plan provides the staff in the Office of Nuclear Material Safety and Safeguards with specific guidance on the review of reclamation plans and license amendments related to reclamation plans. The reclamation plan, submitted by an applicant (in the case of a new application) or a licensee (in the case of an amendment to a previously approved reclamation plan or termination of an existing license) should demonstrate compliance with the applicable criteria in Appendix A to 10 CFR Part 40. The introduction to 10 CFR Part 40 specifically states that "In many cases, flexibility is provided in the criteria to allow achieving an optimum tailings disposal program on a site-specific basis. However, in such cases the objectives, technical alternatives and concerns which must be taken into account in developing a tailings program are identified." The principal purpose of the standard review plan is to present guidance to the NRC staff to ensure a consistent quality and uniformity in NRC reviews of reclamation plans. Each section in this standard review plan contains guidance on what is to be reviewed, the basis for the review, how the staff review is to be done, what the staff will find acceptable in a demonstration of compliance with the regulations, and the conclusions that are sought regarding compliance with the regulations in 10 CFR Part 40. This standard review plan is intended to cover only those aspects of the NRC regulatory mission related to the reclamation of mill tailings sites, including soil and ground-water cleanup, at conventional uranium mills. As such, the standard review plan helps focus the staff review on determining if a tailings impoundment can be constructed, operated, and reclaimed in compliance with the applicable NRC regulations. The standard review plan is also intended to make information about regulatory matters widely available to improve communication, and to help interested members of the public and the uranium recovery industry gain a better understanding of the staff review process. In any of these reviews, the staff will consider licensee-proposed alternatives to Appendix A criteria as described in the Introduction in Appendix A to

10 CFR Part 40. The review would cover the level of protection to the public health and safety and the environment and the level of stabilization and containment of the site. All site-specific licensing decisions based on Appendix A criteria or proposed alternatives will consider the risk to health and safety and the environment and the economic costs involved. Staff guidance for review of environmental reports and preparing environmental assessments is found in NUREG–1748 (NRC, 2001).

For license amendments, the review should focus on the changes proposed in the amendment [see NUREG–1748 (NRC, 2001) for guidance on reviewing historical aspects of site performance]. Reviewers should not review previously accepted actions if they are not part of the proposed amendment, unless the review of the amendment package identifies an impact on previously accepted actions.

For changes to previously approved reclamation plans, the licensee need only submit information pertinent to the proposed change. The licensee need not resubmit a complete reclamation plan covering all aspects of site reclamation, but should present information on the proposed changes to the previously approved plan and its updates as identified in the current NRC license. Reviewers should also analyze the inspection history and operation of the site to see if any major problems have been identified over the course of the license term that would have an effect on reclamation. The operating history of the facility is often a valuable source of information concerning the adequacy of site characterization, the acceptability of radiation protection and monitoring programs, and the sufficiency of other data that may influence staff determination of compliance. NUREG–1757, Volume I, Section 16.2 (NRC, 2002) presents guidance for review of these historical aspects of facility performance. If the changes are found to be acceptable, the license is then amended to identify the revised reclamation plan as the required design for reclamation.

License termination usually involves a confirmation that all applicable reclamation requirements have been met. This includes ensuring completion of stabilization work for the tailings consistent with the accepted reclamation plan and a determination that the licensee has complied with all standards applicable to land structures, and ground-water cleanup. As such, the information in this review plan will be used to help make the necessary conclusions concerning license termination. The four aspects of license termination addressed in this review plan included (i) mill decommissioning, decontamination and disposal; (ii) surface soil cleanup and post cleanup verification; (iii) mill tailings surface stabilization; and (iv) ground water corrective action. Compliance with these four aspects of reclamation, taken together, forms the basis for the staff finding that the design and ground-water cleanup program meet applicable requirements, and that the design and cleanup program have been acceptably completed at the sites and that the licensee has, therefore, met the applicable requirements.

The staff will prepare the following reports to document the review: a technical evaluation report and an environmental assessment. The guidance in NUREG–1748 (NRC, 2001) will be used to prepare the environmental assessment. The provisions of 10 CFR 51.21 require preparation of an environmental assessment unless: (i) the staff finds, based on the environmental assessment, that NRC needs to prepare an environmental impact statement; (ii) another federal agency also involved in the action as a cooperating agency needs to prepare an environmental impact statement; (iii) the effects on the quality of the human environment are likely to be highly controversial; or (iv) 10 CFR 51.22 categorically excludes the necessity to prepare an environmental assessment. Applications for new mills require NRC to prepare an environmental impact statement in accordance with 10 CFR 51.20(b)(18). This standard review plan is intended to guide the preparation of the technical evaluation report.

It is important to note that the acceptance criteria noted in this standard review plan are for the guidance of the Office of Nuclear Material Safety and Safeguards staff responsible for the review of license applications. Review plans are not substitutes for the Commission's regulations, and compliance with a particular standard review plan is not required. Methods and solutions different from those set out in the standard review plan may be acceptable if they provide a basis for the findings requisite to the issuance or continuance of a license by NRC. Use of this standard review plan does not obviate the need for professional judgement; it helps assure overall completeness and uniformity of the staff review.

GENERAL REVIEW PROCEDURE

A licensing review is not intended to be a detailed evaluation of all aspects of facility operations. Specific information about implementation of a program or construction of a design outlined in an application is obtained through the NRC review of procedures and operations done as part of the inspection function. However, some procedures may be required during review of a reclamation or decommissioning plan. The differences between licensing reviews and inspections are shown in Figure 1. For a new license application, the staff will review the proposed reclamation plan and ground water protection program for compliance with the criteria in Appendix A to 10 CFR Part 40. For a license renewal or an amendment to an existing license, the staff will only review proposed changes to the NRC-approved reclamation plan for compliance with criteria in Appendix A to 10 CFR Part 40. If the changes proposed have an adverse impact on the performance or functionality of some of the approved features at the site, then the staff will review those items for their compliance with regulations.

In the case of an amendment application concerning confirmation of site or ground-water cleanup or completion of construction, the reviewer will focus on ensuring that the applicable activities have been completed consistent with the approved review plan. Reviewers will not revisit accepted designs or plans unless the as-completed activity presents problems, such as degradation or reconformation.

Changes to existing licensed activities and conditions require the issuance of an appropriate license amendment. An application for such an amendment should describe the proposed changes in detail and should discuss the potential environmental and health and safety impacts. Amendment requests should be reviewed using the appropriate sections of this document for guidance. NUREG–1757, Volume I (NRC, 2002), contains guidance for examining the historical aspects of facility operations in connection with amendment reviews. The steps of the reclamation plan review are described in the paragraphs that follow.

Acceptance Review

The staff will conduct an acceptance review of a new reclamation plan or changes to a previously approved plan to determine the completeness of the information submitted. The reclamation plan will be considered acceptable for docketing if the information in it is sufficiently complete to initiate a detailed technical review, and reflects an adequate reconnaissance and physical examination of the regional and site conditions, and contains appropriate analyses and design information to demonstrate that the applicable regulatory criteria will be met. Completeness of the environmental report will be determined using the information requirements in 10 CFR 51.45 and the guidance in NUREG–1748 (NRC, 2001). The staff should complete the acceptance review and transmit the results to the applicant within 30 days of the receipt of the application, along with a projected schedule for the remainder of the review. In this transmittal, the staff should note any additional information needed to make the reclamation plan or environmental report complete. Detailed technical questions, although not required, can be included, if they are identified during the acceptance review. If the contents of the reclamation plan or environmental report do not clearly demonstrate compliance with applicable regulatory criteria, then the staff may decline to docket the reclamation plan and will return it to the licensee for revisions.

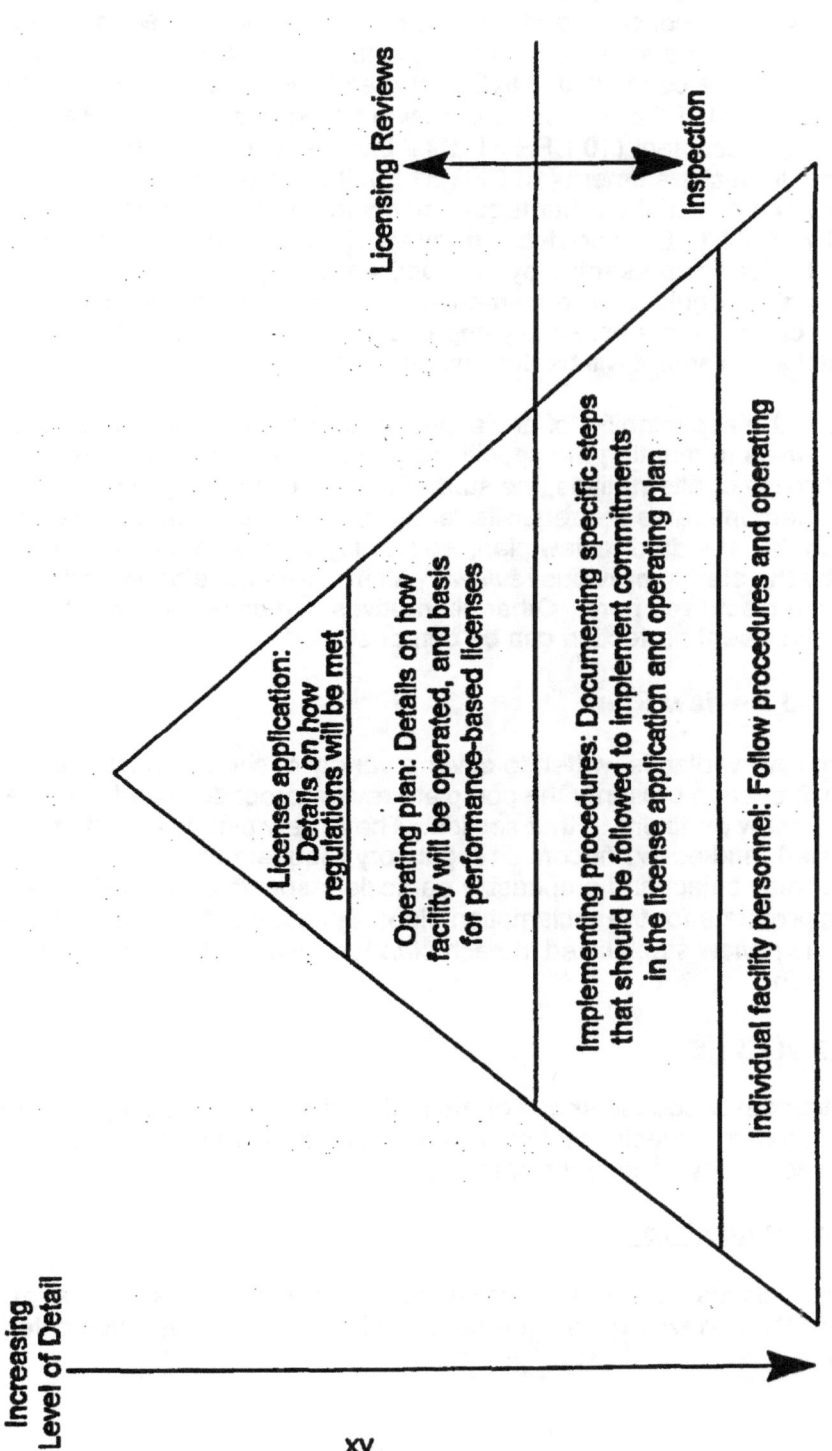

Increasing Level of Detail

Licensing Reviews

Inspection

License application: Details on how regulations will be met

Operating plan: Details on how facility will be operated, and basis for performance-based licenses

Implementing procedures: Documenting specific steps that should be followed to implement commitments in the license application and operating plan

Individual facility personnel: Follow procedures and operating

xv

Figure 1. Schematic of NRC Licensing and Inspection Process and Applicability to Different License Documents

Detailed Review

Following completion of the acceptance review, the staff will conduct a detailed technical review of the reclamation plan. During the detailed review, if there is a need for additional information, the staff will send to the licensee a request for additional information identifying the issue or concern, basis for the concern, and the kind of information needed to resolve the concern. After the staff receives a satisfactory response to the request for additional information, the detailed review will be concluded. NRC documents the results of this review and the basis for acceptance or denial of the requested licensing action in a technical evaluation report, and in an environmental assessment (10 CFR 51.21) if there is a finding of no significant impact, or in an environmental impact statement (10 CFR 51.20) if the reclamation plan is part of an application for a new mill or if one of the other requirements for an environmental impact statement have been met (10 CFR 51.20). The detailed review should evaluate the environmental, economic, and technical evidence presented by the applicant to support the ability of the proposed facility to meet applicable regulatory requirements. In the case of amendments to an existing license as a result of changes to a previously approved reclamation plan, the need for an environmental assessment will be determined on a case-by-case basis.

In determining the acceptability of any aspect of tailings reclamation, the staff will evaluate the use of alternatives to meeting the specific requirements in 10 CFR Part 40, Appendix A. In evaluating the use of alternatives, the staff will determine if the proposed reclamation design satisfactorily demonstrates the requisite requirements of economic benefit and equivalent protection. In this standard review plan, we identify alternatives that have been found to be acceptable by the staff in previous reviews. Alternatives developed by licensees need not be limited to those discussed here. Other alternatives can be proposed, as long as the economic benefit and equivalent protection can be demonstrated.

The Standard Review Plan

The standard review plan is written to cover a variety of site conditions and reclamation designs. Each section presents the complete review procedure and acceptance criteria for all the areas of review pertinent to that section. The review plan is intended as general guidance to the NRC staff, and does not contain regulatory requirements. For any given application, the staff reviewer may select and emphasize particular aspects of each standard review plan section as appropriate for the reclamation plan. Because of this, the staff may not carry out, in detail, all of the review steps listed in each standard review plan section, in the review of every reclamation plan.

I.   Areas of Review

This subsection describes the scope of the review (i.e., what is being reviewed). It contains a brief description of the specific technical information and analyses in the reclamation plan that need to be reviewed by each technical reviewer.

II.   Review Procedures

This subsection discusses the appropriate review technique. It is generally a step-by-step procedure that the reviewer uses to determine whether the acceptance criteria have been met.

III. Acceptance Criteria

This subsection delineates criteria that the reviewer can apply to determine the acceptability of the applicant's compliance demonstration. Although acceptance criteria are not regulatory requirements, the technical bases for these criteria have been derived from 10 CFR Parts 20, 40, and 51, NRC regulatory guides, general design criteria, codes and standards, NRC branch technical positions, standard testing methods (e.g., American Society for Testing and Materials standards), technical papers, and other similar sources. These sources typically contain solutions and approaches previously determined by the staff to be acceptable for making compliance determinations for the specific area of review. These acceptance criteria have been defined so that staff reviewers can use consistent and well-documented approaches for review of all reclamation plans. In the absence of well-defined acceptance criteria, the staff will rely on "professional judgment" and what is normally practiced in the profession. Licensees may take approaches to demonstrating compliance that are different from those in this standard review plan. However, they should recognize that, as is the case for regulatory guides, substantial staff time and effort have gone into the development of these procedures and criteria, and a corresponding amount of time and effort may be required to review and accept new or different solutions and approaches. Thus, licensee-proposed solutions and approaches to safety problems or safety-related design areas other than those described in this standard review plan may require longer review times and NRC requests for more extensive supporting information. The staff is willing to consider proposals for other solutions and approaches on a generic basis, apart from a specific review, to avoid the impact of the additional review time for individual cases.

IV. Evaluation Findings

This subsection presents the staff's general conclusions and findings that result from review of each area of the reclamation plan, as well as identification of the applicable regulatory requirements. Conclusions and findings for a specific site and review area are dependent on the site characteristics and type of licensing action being considered. For each standard review plan section, a conclusion is included in the technical evaluation report/safety evaluation report or in the environmental assessment/environmental impact statement, in which results of the review are published. These documents contain a description of the review; the basis for the staff findings, including aspects of the review selected or emphasized; where the reclamation design or the licensee's plans deviate from the criteria stated in the standard review plan; and the evaluation findings.

Standard Review Plan Updates

The standard review plan will be revised and updated periodically as the need arises to clarify the content or correct errors and to incorporate modifications approved by NRC management.

References

NRC. NUREG–1757, "Consolidated NMSS Decommissioning Guidance." Vol. I. Washington, DC: NRC. January 2002.

NRC Draft NUREG–1748, "Environmental Review Guidance for Licensing Actions Associated with NMSS Programs." Washington, DC: NRC, Office of Nuclear Material Safety and Safeguards. 2001.

1.0 GEOLOGY AND SEISMOLOGY

The reclamation plan and its supporting documents must contain sufficient regional and site-specific geologic and seismologic information related to the proposed disposal site and reclamation design, including regional and site-specific stratigraphy, structure, geomorphology, and seismology. This standard review plan establishes the requirements for staff of the U.S. Nuclear Regulatory Commission (NRC) to conduct and document the review of new reclamation plans for mill tailings impoundments, or amendments to previously approved reclamation plans in the areas of geology and seismology.

1.1 Stratigraphic Features

1.1.1 Areas of Review

The staff should review information presented in the reclamation plan on the regional and site-specific stratigraphy and geology. The reclamation plans should describe surface and subsurface strata and the interpretation of their orientation, occurrence, thickness, composition, age, depositional environment, and interrelationships. The reviewer should coordinate the stratigraphic information with the evaluation of the site's geotechnical stability, surface water and erosion protection, and ground-water resources protection information as described in standard review plan Chapters 2.0, 3.0, and 4.0, respectively. The purpose of this review is to determine if there has been an acceptable characterization of site and regional stratigraphy so that sufficient information has been presented for use in the reclamation plan and design of the tailings cell.

1.1.2 Review Procedures

The reviewer should examine the description and discussion of the regional and site-specific features to determine if a thorough evaluation of the regional and site stratigraphy has been presented.

The following specific descriptive information should be reviewed to determine its adequacy for characterizing the regional and site-specific stratigraphic features:

(1)   Description of regional stratigraphic units by rock classification and type.

(2)   Distribution of regional stratigraphic units.

(3)   Age relationships of regional and site-specific stratigraphic units.

(4)   Detailed site stratigraphy based on outcrop and well borings conducted to determine rock types and their texture, composition, distribution, thickness, and environment of deposition.

The staff determination of compliance should be based in part on professional judgment, considering the complexity of the subsurface conditions at the site.

Geology and Seismology

1.1.3 Acceptance Criteria

The characterization of regional and site stratigraphy will be acceptable if the information presented conforms to the following criteria:

(1)   The regional and site-specific stratigraphy are described in sufficient detail to produce an adequate understanding of the site-specific subsurface characteristics, including descriptions of major stratigraphic units and their orientations, age relationships, thicknesses, environments of deposition distributions, and any stratigraphic features (e.g., facies changes) likely to affect site stability or ground-water resource protection.

(2)   Stratigraphic units are described in sufficient detail to provide input to a geotechnical stability analysis.

(3)   Descriptions of regional and site-specific stratigraphic units contain sufficient information for input to an analysis of ground water resources and the protection thereof.

(4)   Regional stratigraphic information is discussed in sufficient detail to support site-specific information.

(5)   Descriptions of the regional and site stratigraphy are based on published literature and site data and conform to standard geological classifications.

(6)   Discussions of regional stratigraphy are adequately referenced and supported by published reports, maps, logs, and cross sections.

(7)   Site descriptions are based on field investigations and adequate sampling to define physical and chemical properties of surface and subsurface materials such as soils and underlying geologic formations at the site.

(8)   Maps are at a scale sufficient to show the locations of all site explorations such as borings, geophysical surveys, trenches, and sample locations.

Where insufficient information is presented to support interpretations and conclusions, the reviewer will request additional investigations or data gathering. Staff determination of compliance should be based in part on professional judgment, considering the complexity of the site conditions.

1.1.4 Evaluation Findings

If the staff review, as described in standard review plan Section 1.1, results in the acceptance of the characterization of regional and site stratigraphy, the following conclusions may be presented in the technical evaluation report.

The staff has completed its review of the characterization of the regional and site stratigraphy at the_____ uranium mill facility. This review included an evaluation using the review procedures in Section 1.1.2 and the acceptance criteria outlined in Section 1.1.3 of this standard review plan.

The licensee has provided an acceptable description of the stratigraphic features by presenting a description of the site and regional stratigraphy using published information and information collected for the specific purpose of supporting determinations of geotechnical stability and ground water analyses at the site. Data gathering, investigations, and analyses have used acceptable standards and practices. Data and interpretations of data are presented to allow effective incorporation into geotechnical and ground-water analyses.

On the basis of the information and analysis presented in the review plan on the stratigraphic features at the _____ uranium mill facility, the NRC staff concludes that the information is sufficient to support a decision with reasonable assurance that the requirements of 10 CFR Part 40, Appendix A, Criterion 4(e) have been met. These require that tailings impoundments not be located near a capable fault that could cause a maximum credible earthquake larger than that which the impoundment could reasonably be expected to withstand , or that an acceptable alternate method of determination of seismic hazard has been used. If a probabilistic seismic hazard analysis is used as an alternate method, the applicant has presented sufficient information to support an analysis of the facility design for the operational and post-operational periods. The description of the physical and chemical properties of the underlying soils and geologic formations of the site is sufficient to meet the requirements of 10 CFR Part 40, Appendix A, Criterion 5G(2) with regard to the extent to which they will control transport of contaminants and solutions. Reasonable assurance has also been provided that the requirements of 10 CFR Part 40, Appendix A, Criterion 6(1), which requires that the design of the disposal facility provide reasonable assurance of control of radiological hazards to be effective for 1,000 years, to the extent reasonably achievable, and, in any case, for at least 200 years, have been met.

1.1.5 References

None.

1.2 Structural and Tectonic Features

1.2.1 Areas of Review

The staff should review information presented in the reclamation plan on the regional and site-specific structural and tectonic setting. The reclamation plan should contain a definition of surface and subsurface structural and tectonic features and an interpretation of their origin, occurrence, age, and potential impacts, if any, on the stability of the site. Review of the structural and tectonic information should be coordinated with the evaluation of the site's geotechnical stability, surface water and erosion protection, and ground-water resources protection information as described in standard review plan Chapters 2.0, 3.0,and 4.0, respectively. The reviewer will determine whether the information presented is sufficient to support an analysis of geologic features as they affect the facility.

1.2.2 Review Procedures

The reviewer should examine the description and discussion of the regional and site-specific information to determine if a thorough evaluation of structural and tectonic features has been presented. This may include analyses of photogrammetric data, results of field reconnaissance

Geology and Seismology

and detailed mapping, review of pertinent literature, and
review of geophysical data and studies. Features that should be considered in the review
include structural features such as faults and fractures, crustal deformation, and volcanic
features that may affect the site stability or ground-water conditions.

The following specific descriptive information should be reviewed to determine its adequacy for
characterizing the regional and site-specific structural features necessary to support the
evaluations of reclamation system performance:

(1)   Description and location of regional structural features based on published information
        and field reconnaissance, including the geologic attitude of key stratigraphic units.

(2)   Description and location of site subsurface structural features from sources such as
        available borings, drill logs, geophysical logs and data, and existing literature.

(3)   Description of any volcanic features such as flows, cones, plugs, or dikes located in the
        site region.

(4)   Age relationships of regional and site-specific structural and tectonic features.

(5)   Discussion of published literature containing interpretations of any of the information in
        previous Items 1, 2, 3, and 4.

Staff determination of compliance should be based in part on professional judgment,
considering the complexity of the subsurface conditions at the site.

1.2.3   Acceptance Criteria

The characterization of regional and site structural features will be acceptable if the information
presented in the reclamation plan conforms to the following criteria:

(1)   Descriptions of regional and site-specific structural and tectonic features are based on
        published literature and gathered data.

(2)   Regional structural and tectonic features, particularly faults, are defined in sufficient
        detail to present an adequate understanding of the structural geologic conditions that
        may have a likelihood of affecting the site stability or ground-water regime.

(3)   Site-specific structural and tectonic features, particularly faults, are described in
        sufficient detail to present adequate information for an analysis of the site stability.
        Information presented adequately addresses the uncertainties and variability within the
        site area and the potential impacts on the disposal facility.

(4)   The structural and tectonic province or provinces that influence the site seismicity are
        identified and described.

(5)   The tectonic history of the pertinent province(s) is discussed in sufficient detail to
        support an analysis of the potential for disruption of the site by tectonic activity.

(6)   Discussions of structural, tectonic, and volcanic features are adequately referenced and are supported by maps, logs, and cross sections showing locations of all site explorations and surveys, and depicting surface and subsurface structural and tectonic features.

(7)   Descriptions contain discussions of age relationships of structural and tectonic features.

Where insufficient information is presented to support interpretations and conclusions, the reviewer will request additional investigations or data gathering. Staff determination of compliance should be based in part on professional judgment, considering the complexity of the site conditions.

1.2.4   Evaluation Findings

If the staff review, as described in standard review plan Section 1.2, results in the acceptance of the characterization of the structural and tectonic features of the region and site, the following conclusions may be presented in the technical evaluation report.

The staff has completed its review of the characterization of structural and tectonic features at the _____ uranium mill facility. This review included an evaluation using the review procedures in Section 1.2.2 and the acceptance criteria outlined in Section 1.2.3 of this standard review plan.

The licensee has acceptably described the regional and site-specific structural and tectonic features by presenting discussions and interpretations of pertinent data and reports that may have an impact on the site or tailings disposal system. Information presented includes descriptions of any faults capable of disrupting the site and any other information necessary to support an analysis of the geotechnical stability or ground-water conditions at the site. In addition, the staff concludes that the licensee has used acceptable methods of investigation and analysis to support its conclusions.

On the basis of the information and analysis presented in the review plan on the structural and tectonic features at the _____ uranium mill facility, the NRC staff concludes that the information is sufficient to support a decision with reasonable assurance that the requirements of 10 CFR Part 40, Appendix A, Criterion 4(e) have been met. These require that tailings impoundments not be located near a capable fault that could cause a maximum credible earthquake larger than that which the impoundment could reasonably be expected to withstand, or that an acceptable alternate method of determination of seismic hazard has been used. If a probabilistic seismic hazard analysis is used as an alternate method, the applicant has presented sufficient information to support an analysis of the facility design for the operational and postoperational periods. Reasonable assurance has also been provided that the requirements of 10 CFR Part 40, Appendix A, Criterion 6(1), which requires that the design of the disposal facility provide reasonable assurance of control of radiological hazards to be effective for 1,000 years, to the extent reasonably achievable, and, in any case, for at least 200 years, have been met.

1.2.5   References

None.

1.3   Geomorphic Features

1.3.1   Areas of Review

The staff should review the information presented in the reclamation plan on the regional and site-specific geomorphic features. The reclamation plan should analyze regional and local landforms to determine evidence for geomorphic processes that may impact the long-term stability of the site, including information to support an evaluation of the potential for any destructive geomorphic processes, such as mass wasting, extreme erosion, and stream encroachment. The reviewer should coordinate the geomorphic information with the evaluation of the site's geotechnical stability and surface water and erosion protection information as described in standard review plan Chapters 2.0 and 3.0, respectively. The results of this review will be used to determine the acceptability of the design during operation and long-term stabilization.

1.3.2   Review Procedures

The reviewer should examine the description and discussion of the regional and site-specific geomorphic information to determine if a thorough evaluation has been presented. Information should be detailed enough for the reviewer to make a determination regarding the geomorphic stability of the site.

The following specific descriptive information should be reviewed to determine the acceptability of the assessment of the regional and site-specific geomorphology as it relates to geomorphic stability of the site:

(1)   Description of the physiographic (geomorphic) province(s) in which the site is located, including a discussion of the distinguishing characteristics such as elevation and relief.

(2)   Discussion of the active processes, such as erosion, mass wasting, and stream encroachment within the site region and the nature and extent of those processes.

(3)   Topographic maps depicting geomorphic surfaces, physiographic provinces, landforms, drainage networks, rivers, surficial geologic units, areas of subsidence, and geomorphic hazards.

(4)   Aerial photographs of the site area.

(5)   Discussion of the age, occurrence, and origin of geomorphic features, in particular those that may adversely affect site stability.

1.3.3   Acceptance Criteria

The characterization of regional and site geomorphic features and geomorphic stability will be acceptable if the information presented conforms to the following criteria:

(1)   Descriptions of the regional and site-specific geomorphology and geomorphic processes include information sufficient to allow the reviewer to assess the nature and extent of major active processes that may modify the present-day topography of the geomorphic province(s) and the site area.

(2)   The geomorphic features, particularly potential geomorphic hazards, are clearly delineated on topographic base maps of adequate scale to enable the reviewer to assess their occurrence and distribution.

(3)   Descriptions are adequately referenced and are supported by published reports and maps or site data.

(4)   The regional and site-specific geomorphology and geomorphic processes are described in sufficient detail to support an analysis of the geomorphic and geotechnical stability of the site.

Where insufficient information is presented to support interpretations and conclusions, the reviewer will request additional investigations or data gathering. Staff determination of compliance should be based in part on professional judgment, considering the complexity of the site conditions.

1.3.4   Evaluation Findings

If the staff review, as described in standard review plan Section 1.3, results in the acceptance of the characterization of the geomorphic features of the region and site and provides information sufficient to support an assessment of the geomorphic stability, the following conclusions may be presented in the technical evaluation report.

The NRC has completed its review of the information concerning the characterization of geomorphic features at the _____ uranium mill facility. This review included an evaluation using the review procedures in Section 1.3.2 and the acceptance criteria outlined in Section 1.3.3. of this standard review plan.

The licensee has acceptably described the geomorphic features by presenting an adequate description of regional and site geomorphology using published information and information collected for the specific purpose of supporting determinations of the stability of site. Data gathering, investigations, and analyses have used acceptable standards and practices. Data and interpretations are presented to allow effective

incorporation into other site analyses.

On the basis of the information and analysis presented in the review plan on the geomorphic features at the _____ uranium mill facility, the NRC staff concludes that the information is sufficient to support a decision with reasonable assurance that the requirements of 10 CFR Part 40, Appendix A, Criterion 6(1), which requires that the design of the disposal facility provide reasonable assurance of control of radiological hazards to be effective for 1,000 years, to the extent reasonably achievable, and, in any case, for at least 200 years, have been met.

Geology and Seismology

1.3.5 References

None.

1.4   Seismicity and Ground Motion Estimates

1.4.1 Areas of Review

The staff should review information presented in the reclamation plan on the regional and site-specific seismicity and the basis for determining the vibratory ground motion (peak horizontal acceleration) at the site from seismic events. The purpose of this review is to determine the potential for seismic events to affect the site. The reviewer will determine whether the information presented is sufficient to support an analysis of the design for the operational and closure periods.

1.4.2 Review Procedures

The reviewer should examine the description and discussion of the regional and site-specific information to determine if a thorough evaluation of the potential for seismic activity has been presented. The information should be sufficient to enable the reviewer to determine the vibratory ground motion (peak horizontal acceleration) at the site from seismic events.

The following specific descriptive information should be reviewed to determine the acceptability of the characterization of the seismicity and the assessment of the stability of the site and geotechnical design:

(1)   A listing of all recorded earthquakes in the tectonic province in which the site is located and in other tectonic provinces within 200 km [124 mi] of the site. This listing should contain the date of occurrence of the earthquake, its magnitude, and the location of the epicenter. Since earthquakes have at times been reported in terms of intensity at a given location, or effect on ground, structures, and people at a specific location, some of this information may have to be

estimated by use of appropriate empirical relationships.

(2)   Data obtained by standard photogeologic analysis and field reconnaissance of the study area and from review of the pertinent literature. Information in the form of maps, papers, or other data, specific to the area or region, generated by state and federal agencies or published in the literature, should be utilized.

(3)   An association of epicenters or locations of highest intensity of historic earthquakes with tectonic structures, where possible. Epicenters or locations of highest intensity that cannot be reasonably identified with tectonic structures should be identified with tectonic provinces.

(4)   Maps on which the locations of epicenters of historic earthquakes, associated tectonic structures, and tectonic provinces have been depicted.

(5)   The applicant proposed maximum earthquakes associated with each tectonic province or capable fault or structure.

(6)   Deterministic and/or probabilistic seismic hazard analyses.

For a deterministic analysis, the potential ground motion at the site from capable faults that might affect the licensed area should be assessed. The term "capable fault" as used in 10 CFR Part 40, Appendix A, Criterion 4(e), has the same meaning as defined in Section III(g) of Appendix A to 10 CFR Part 100. Alternatively, the licensee may choose to use the term "capable tectonic source" as defined in Appendix A to Regulatory Guide 1.165 (NRC, 1997) to conduct its analysis.

A probabilistic seismic hazard analysis yields a curve of exceedence probability versus peak horizontal acceleration. The 10^{-4} value represents a 1 in 10 chance of the site exceeding the peak horizontal acceleration in a 1,000-year period, which is appropriate for a 1,000-year design life. The seismic hazard analysis of uranium recovery mill sites by Bernreuter, et al. (1994), contains probabilistic analyses for Title II mill sites. The study by Bernreuter, et al. (1994) is intended as a screening study; the probabilistic seismic hazard estimates are not site specific and are only calculated for random earthquakes.

(7)   Seismic design ground motion (peak horizontal acceleration).

Staff determination of compliance should be based in part on professional judgment, considering the complexity of the regional and site-specific seismicity. The reviewer will focus on evaluating the maximum credible earthquake, as required by 10 CFR Part 40, Appendix A, Criterion 4(e), unless an alternate

method of determining ground motion is presented as allowed in the Introduction to Appendix A. One such alternative to the maximum credible earthquake is a probabilistic seismic hazard analysis, which is presented in Section 1.4.3, below.

1.4.3   Acceptance Criteria

The regional and site-specific seismicity and ground motion estimates will be acceptable if the following criteria are met:

(1)   The information presented on the regional and site-specific seismicity contains sufficient detail to allow the staff to determine the vibratory ground motion (peak horizontal acceleration) at the site caused by seismic events and to further use that determination to assess the geotechnical stability of the site. The geotechnical stability of the site is sufficient to control radiological hazards for 1,000 years to the extent reasonably achievable, and, in any case, for at least 200 years.

(2)   In conducting this review, the staff will consider a deterministic and/or a probabilistic seismic hazard analysis as an acceptable method for selecting the peak horizontal acceleration for a site. An analysis of the geotechnical stability of the design proposed in the reclamation plan will be based on the resultant peak horizontal acceleration

Geology and Seismology

(Chapter 2.0, "Geotechnical Stability," of this standard review plan).

(a)   Deterministic Analysis: The use of a deterministic seismic hazard analysis is acceptable if:

   (i)   Capability is determined by suitable methods, such as those outlined by Slemmons (1977).

   (ii)   Fault length versus magnitude relationships for determining the maximum magnitude earthquake that may be produced by each capable fault or capable tectonic source are developed using acceptable approaches such as those of Slemmons, et al. (1982); Bonilla, et al. (1984); or Wells and Coppersmith (1994).

   (iii) For each maximum magnitude earthquake, the peak horizontal acceleration at the site is determined using the applicable attenuation relationship between earthquake magnitude and distance for the site. Campbell (1997); Campbell and Bozorgnia (1994); and Boore, et al. (1993, 1997) offer examples of acceptable attenuation relationships. In applying the relationship, the site-to-source distance should be the distance between the site and the

      closest approach of the fault.

   (iv)   The peak horizontal acceleration value adopted for each capable fault or tectonic source is not less than the median value provided by the attenuation relationship. Possible soil amplification effects are considered.

   (v)   To assess potential ground motion at the site from earthquakes not associated with known tectonic structures (i.e., random or floating earthquakes), the largest floating earthquakes reasonably expected within the tectonic province are identified. In addition, the largest floating earthquakes characteristic of any adjacent tectonic provinces are identified, if such earthquakes cause appreciable ground motion at the site. For each of these earthquakes, the peak horizontal acceleration at the site is calculated as stated previously, with 15 km [9 mi] used as the site-to-source distance for floating earthquakes within the host tectonic province. For floating earthquakes in other tectonic provinces, the distance between the site and the closest approach of the province boundary is used as the site-to-source distance.

   (vi)   The peak horizontal acceleration for the site is the maximum value of the peak horizontal accelerations determined for earthquakes from all capable faults, tectonic sources, and tectonic provinces.

(b)   Probabilistic Analysis: The use of a probabilistic seismic hazard analysis as an alternative to the requirements of 10 CFR Part 40, Appendix A, Criterion 4(e), is acceptable, as is stated in the Introduction to Appendix A, if:

(i)   It is shown that the design proposed by the licensee will achieve a level of stabilization and containment, and a level of protection for public health and safety and the environment, which is equivalent to, to the extent practicable, or more stringent than that achieved by the requirements of 10 CFR Part 40, Appendix A.

(ii)   The licensee takes into account local conditions when estimating the seismic design of the facility because peak horizontal acceleration values are often calculated for hypothetical rock foundations. The effects of local site conditions on the peak ground acceleration are reviewed in Chapter 2.0 in the standard review plan.

(3)   The presentation on seismotectonic stability is acceptable if sufficient information is presented to support interpretations and conclusions. If the staff should conclude that the information presented is insufficient, it will request additional information or investigations. Staff determination of compliance should be based, in part, on professional judgment, considering the complexity of site and seismic conditions.

1.4.4 Evaluation Findings

If the staff review as described in standard review plan Section 1.4 results in the acceptance of the characterization of the seismicity of the region and site and the seismic design ground motion, the following conclusions may be presented in the technical evaluation report.

The staff has completed its review of the characterization of the seismicity at the _____ uranium mill facility. This review included an evaluation using the review procedures in Section 1.4.2 and the acceptance criteria outlined in Section 1.4.3 of this standard review plan.

The licensee has presented information and investigations that support its conclusions about the seismic characterization of the site and the seismic design value. Information presented includes descriptions of historical earthquakes, locations of their epicenters, an analysis of the seismic hazard at the site, and the design peak horizontal acceleration. The staff concludes that the information presented is sufficient to support an analysis of the geotechnical stability. In addition, the staff concludes that the licensee has used acceptable methods of investigation and analysis to support its conclusions.

On the basis of the information and analysis presented in the review plan on the seismicity and ground motion estimates at the _____ uranium mill facility, the NRC staff concludes that the information is sufficient to support a decision with reasonable assurance that the requirements of 10 CFR Part 40, Appendix A, Criterion 4(e), have been met. These require that tailings impoundments not be located near a capable fault that would cause a maximum credible earthquake larger than that which the impoundment could reasonably be expected to withstand, or that an acceptable alternate method of determination of seismic hazard has been used. If a probabilistic seismic hazard analysis is used as an alternate method, the applicant has presented sufficient information to support an analysis of the facility design for the operational and postoperational periods. Reasonable assurance has also been provided that the requirements of 10 CFR Part 40, Appendix A, Criterion 6(1), which requires that the design

Geology and Seismology

of the disposal facility provide reasonable assurance of control of radiological hazards to be effective for 1,000 years, to the extent reasonably achievable, and, in any case, for at least 200 years, have been met.

1.4.5　References

Bernreuter, D., E. McDermott, and J. Wagoner. "Seismic Hazard Analysis of Title II Reclamation Plans." Livermore, California: Lawrence Livermore National Laboratory. 1994.

Bonilla, M.G., R. K. Mark, and J.J. Lienkaemper. "Statistical Relations Among Earthquake Magnitude, Surface Rupture Length, and Surface Fault Displacement." *Bulletin of the Seismological Society of America*. Vol. 74. pp. 2,379–2,411. 1984.

Boore, D.M., W.B. Joyner, and T.E. Fumal. "Estimation of Response Spectra and Peak Acceleration From Western North American Earthquakes: An Interim Report." Open-File Report 93-509. U.S. Geological Survey. 1993.

————. "Equations for Estimating Horizontal Response Spectra and Peak Acceleration from Western North American Earthquakes: A Summary of Recent Work." *Seismological Research Letters*. Vol. 68. pp. 128,153. 1997.

Campbell, K. "Empirical Near-Source Attenuation Relationships for Horizontal and Vertical Components of Peak Ground Acceleration, Peak Velocity, and Pseudo-Absolute Acceleration Response Spectra." *Seismological Research Letters*. Vol. 68. pp. 154–179. 1997.

Campbell, K.W. and Y. Bozorgnia. "Near Source Attenuation of Peak Horizontal Acceleration From Worldwide Accelerograms Recorded From 1975 to 1993." Fifth U.S. National Conference on Earthquake Engineering, Chicago, Illinois, July 10–14. 1994.

NRC. Regulatory Guide 1.165, "Identification and Characterization of Seismic Sources and Determination of Safe Shutdown Earthquake Ground Motion." Washington, DC: NRC, Office of Standard Development. March 1997.

Slemmons, D.B. "State-of-the-Art for Assessing Earthquake Hazards in the United States: Report 6, Faults and Earthquake Magnitudes." Miscellaneous Paper S-73-1. Vicksburg, Mississippi: U.S. Corps of Engineers, U.S. Army Engineer Waterways Experiment Station. 1977.

Slemmons, D.B., P. O'Malley, R.A. Whitney, D.H. Chung, and D.L. Bernreuter. "Assessment of Active Faults for Maximum Credible Earthquakes of the Southern California-Northern Baja Region." Publication No. UCID 19125 University of California. Livermore, California: Lawrence Livermore National Laboratory. 1982.

Wells, D.L. and K.J. Coppersmith. "New Empirical Relationships Among Magnitude, Rupture Length, Rupture Width, Rupture Area, and Surface Displacement." *Bulletin of the Seismological Society of America*. Vol. 84. pp. 974–1,002. 1994.

2.0 GEOTECHNICAL STABILITY

The reclamation plan and its supporting documents must contain geotechnical information, design details, and construction considerations related to the proposed disposal site and to all materials associated with the reclamation design, including soil and rock cover, foundation materials, contaminated materials, and other materials, for any zones (liners, filters, or capillary breaks). Standard review plan Chapter 2.0 establishes the procedures for NRC staff to conduct and document the review of geotechnical stability aspects of reclamation plans for mill tailings impoundments, amendments to the approved reclamation plans, or license termination.

2.1 Site and Uranium Mill Tailings Characteristics

2.1.1 Areas of Review

The staff should review information presented in the reclamation plan on the geotechnical aspects of the regional and site geology and stratigraphy, the geotechnical characteristics of the uranium mill tailings and other materials designated for stabilization, and borrow area material characteristics. "Other materials" are contaminated soil from site cleanup operations, tailings from other sites accepted for disposal at this site, and any contaminated materials from mill decommissioning activities to be disposed of at this site. This review should cover exploration data, sampling and laboratory techniques, test results, descriptions of physical properties, and static and dynamic geotechnical engineering parameters of the materials, as well as discussions of ground-water conditions (e.g., perched, confined, or unconfined) for all critical subsurface strata at the site, including information on the fluctuations of the hydraulic head. Review of the ground-water information should be coordinated with the review of information on ground-water resources protection, as described in standard review plan Chapter 4.0. Review of geologic, stratigraphic, and seismologic information should be coordinated with the review of the geology and seismology information as described in standard review plan Chapter 1.0. Borrow area restoration plans should be evaluated.

2.1.2 Review Procedures

The information to be reviewed depends on whether the proposed tailings disposal is below grade, either in mines or specially excavated pits, or in above ground impoundments. The reviewer should focus on the appropriateness of the site characterization for the proposed tailings disposal scheme. The reviewer should examine the site stratigraphy and evaluation of engineering properties of the underlying materials at the site, uranium mill tailings, other materials, and borrow materials to determine if appropriate methods were properly used in characterizing the materials.

The reviewer should examine the following specific descriptive information to determine its adequacy for characterizing the site and for supporting the evaluations of reclamation system performance:

(1)   Site stratigraphy, based on borings and other investigations conducted to determine the type, location, and thickness of underlying materials.

Geotechnical Stability

(2)   Regional and site-specific seismologic information to determine the potential for impact on the geotechnical stability of the site and site structures.

(3)   Stratigraphy specifying type, location, and thickness of borrow material and other materials designated for stabilization in the tailings disposal cell.

(4)   *In situ* testing programs and procedures conducted to determine the engineering properties of underlying materials at the site, borrow area material, other materials, and tailings.

(5)   Sampling programs conducted to obtain laboratory samples for determination of engineering properties of borrow materials, underlying materials at the site, other materials, and tailings.

(6)   Laboratory testing used to determine the engineering properties of borrow materials, underlying materials at the site, other materials, and tailings.

(7)   Physical and engineering properties of borrow materials, underlying materials at the site, other materials, and tailings.

(8)   Records of historical ground-water-level fluctuations at the site, to the extent that they are available.

The reviewer should evaluate methods used to characterize the site to ensure that they comply with generally accepted standards, such as those of the American Society for Testing and Materials and those which are commonly used in the geotechnical engineering profession. Areas to be examined in this respect include the *in situ* and laboratory testing programs, sampling techniques, and analyses for determining the physical and engineering properties of materials at the site. Field investigations and laboratory testing procedures not commonly used in the geotechnical engineering profession will be reviewed in detail.

Staff determination of compliance should be based in part on professional judgment, considering the complexity of the site subsurface conditions.

2.1.3   Acceptance Criteria

The site characterization information constitutes part of the input data needed for analysis and design of the tailings impoundment facility. The site characterization will be acceptable if it provides the needed input for the design and analysis of the disposal facility and meets the following criteria:

(1)   The site stratigraphy is described in sufficient detail to provide an understanding of the site-specific subsurface features, including structural features and other characteristics of underlying soil and rock.

(2)   Information on regional and local faults and seismicity, as obtained from field data, published literature, and historical records is presented in sufficient detail to effectively incorporate that information into a geotechnical stability analyses. (Note: This aspect of the review should be coordinated with the geology and seismology review performed in accordance with standard review plan Chapter 1.)

(3)   Sampling scope and techniques are appropriate and sufficient to ensure that samples collected are representative of the range of *in situ* soil conditions, taking into consideration variability and uncertainties in such conditions within the site.

(4)   For all soils that might be unstable because of their physical or chemical properties, locations and dimensions are identified and the properties have been documented.

(5)   Investigations (including laboratory and field testing) are conducted using appropriate standards published by the American Society for Testing and Materials or the International Society for Rock Mechanics and are sufficient to establish the static and dynamic engineering parameters of borrow materials, other materials, tailings, and underlying soil and rock materials at the site (NRC, 1978, 1979).

(6)   A detailed discussion of laboratory sample preparation techniques is presented, when standard procedures are not used.

        For critical laboratory tests, details such as how saturation of the sample was determined and maintained during testing, or how the pore pressures changed are provided. A detailed and quantitative discussion of the criteria used to verify that the samples were properly taken and tested in sufficient number to define the critical soil parameters for the site is presented. In the case of tailings material (e.g., license amendment reviews), the evaluations of its strength and settlement characteristics are presented in detail.

(7)   Parameter values are presented to enable evaluation of properties of mill tailings, borrow materials, other materials, and underlying soil and rock, including the following:

        (a)   Compressibility and rate of consolidation

        (b)   Shear strength, including, for sensitive soils, possible loss of shear strength resulting from strain-softening

        (c)   Liquefaction potential

        (d)   Permeability

        (e)   Dispersion characteristics

        (f)   Swelling and shrinkage

Geotechnical Stability

        (g)   Long-term moisture content for radon barrier material

        (h)   Cover cracking

(8)   Soil stratigraphy and relevant parameters that are used in the geotechnical evaluations (settlement, stability, liquefaction potential, etc.) are discussed in detail.

(9)   Records of historical ground-water-level fluctuations at the site as obtained from monitoring local wells and springs and/or by analysis of piezometer and permeability data from tests conducted at the site are presented in sufficient detail to effectively incorporate the information into geotechnical stability analyses. (Note: This aspect of the review should be coordinated with the hydrogeologic characterization review performed according to standard review plan Chapter 4.0.)

The information should be sufficient to provide the required input for the design of the facility and to enable the reviewer to assess compliance with the regulatory requirements, such as site features contributing to waste isolation; facility location with respect to an active fault; and reasonable assurance of control of radiological hazards to be effective for 1,000 years to the extent reasonably achievable, and in any case, for at least 200 years.

2.1.4   Evaluation Findings

If the staff review as described in standard review plan Section 2.1 results in the acceptance of the characterization of the site and uranium mill tailings sufficient to support a conclusion regarding the geotechnical stability of the site, the following conclusions may be presented in the technical evaluation report:

The staff has completed its review of the geotechnical characteristics of the site and uranium mill tailings at the _____ uranium mill facility. This review included an evaluation using the review procedures in Section 2.1.2 and the acceptance criteria outlined in Section 2.1.3 of this standard review plan.

The licensee has acceptably described the geotechnical characteristics of the site and uranium mill tailings based on sampling techniques that are acceptable, and will ensure that a representative range of *in situ* soil conditions will be examined. Unstable soils have been identified. Investigations and analyses have used acceptable standards and practices. Laboratory sample preparation and testing techniques are appropriately described and include: (1) compressibility and rate of consolidation, (2) shear strength, (3) liquefaction potential, (4) permeability, (5) dispersion characteristics, (6) swelling and shrinkage, and (7) physical properties. Records of historic ground-water-level fluctuations are presented to allow effective incorporation into geotechnical stability analyses.

On the basis of the information presented in the application and the detailed review conducted of geotechnical the characteristics of the site and uranium mill tailings at the _____ uranium mill facility, the NRC staff concludes that the geotechnical characterization of the site and uranium mill tailings and associated conceptual and numerical models provide an acceptable input which, along with other information such as results of design analysis, will enable the staff to make a finding on the demonstration of compliance with the following criteria in Appendix A to 10 CFR Part 40: (1) Criterion 1, which relates to the site features that contribute to the permanent waste isolation characteristics of the site; (2) Criterion 3, which states the primary option for disposal of tailings is placement below grade, either in mines or specially excavated pits (if applicable for the site); (3) Criterion 4(e), which requires that the impoundment not be located near a capable fault on which a maximum credible earthquake, larger than one that the impoundment could reasonably be expected to withstand, might occur; (4) Criterion 5(G)(2), relating to the permeability characteristics of the site; and (5) Criterion 6(1), which requires reasonable assurance of control of radiological hazards to be effective for 1,000 years to the extent reasonably achievable, and in any case for at least 200 years.

2.1.5 References

American Society for Testing and Materials Standards:

D 420, "Guide for Investigating and Sampling Soil and Rock."

D 421, "Practice for Dry Preparation of Soil Samples for Particle-Size Analysis and Determination of Soil Constants."

D 422, "Method for Particle-Size Analysis of Soils."

D 653, "Terminology Relating to Soil, Rock, and Contained Fluids."

D 854, "Test Method for Specific Gravity of Soils."

D 1140, "Test Method for Amount of Material in Soils Finer Than the No. 200 Sieve."

D 1452, "Practice for Soil Investigation and Sampling by Auger Borings."

D 1586, "Method for Penetration Test and Split-Barrel Sampling of Soils."

D 1587, "Practice for Thin-Walled Tube Sampling of Soils."

D 2113, "Practice for Diamond Core Drilling for Site Investigation."

D 2166, "Test Method for Unconfined Compressive Strength of Cohesive Soil."

D 2216, "Method for Laboratory Determination of Water (Moisture) Content of Soil, Rock and Soil-Aggregate Mixtures."

D 2217, "Practice for Wet Preparation of Soil Samples for Particle-Size Analysis and Determination of Soil Constants."

D 2487, "Test Method for Classification of Soils for Engineering Purposes."

D 2488, "Practice for Description and Identification of Soils (Visual-Manual Procedure)."

D 2573, "Test Method for Field Vane Shear Test in Cohesive Soils."

D 3441, "Method for Deep, Quasi-Static, Cone and Friction-Cone Penetration Tests of Soil."

D 3550, "Practice for Ring-Lined Barrel Sampling of Soils."

D 4221, "Test Method for Dispersive Characteristics of Clay Soil by Double Hydrometer."

D 4318, "Test Method for Liquid Limit, Plastic Limit, and Plasticity Index of Soils."

D 4647, "Test Method for Identification and Classification of Dispersive Clay Soils by the Pinhole Test."

D 4750, "Test Method for Determining Subsurface Liquid Levels in a Borehole or Monitoring Well (Observation Well)."

NRC. Regulatory Guide 1.132, "Site Investigations for Foundations of Nuclear Power Plants." Rev. 1. Washington, DC: NRC, Office of Standards Development. March 1979.

NRC Regulatory Guide 1.138, "Laboratory Investigations of Soils for Engineering Analysis and Design of Nuclear Power Plants." Washington, DC: NRC, Office of Standards Development. April 1978.

2.2   Slope Stability

2.2.1   Areas of Review

The staff should examine exploration data, test results, slope characterization data, design details, and static and dynamic analyses related to the stability of all natural and manmade earth and rock slopes whose failure, under any of the conditions to which they could be exposed throughout the period of regulatory interest, could adversely affect the integrity of the slopes or embankments. This review should also include examination of static and dynamic materials properties, test and design methods, pore pressures within and beneath the embankment, and the design seismic coefficient. Information on the design seismic event should be obtained from results of the review completed using standard review plan Chapter 1.0. The review will focus on (i) the design of the impoundment during operation when a large volume of tailings liquor would be present and (ii) the stability of the impoundment over the long term.

2.2.2 Review Procedures

The reviewer should examine data gathered from site investigations, such as borings: maps; laboratory and field tests; soil profiles; site plans; results of seismic investigations; permeability tests; and static, dynamic, or pseudostatic stability analyses to determine whether the assumptions and analyses used in the reclamation plan are conservative. The degree of conservatism required depends on the type of analysis used, the variability and uncertainty in the values of the parameters considered in the slope stability analysis, the number of borings, the sampling program, the extent of the laboratory testing program, and the resultant safety factor. For instances in which safety factors are low, the reviewer should ensure that reasonable ranges of soil properties have been considered. Other factors, such as flood conditions, pore pressure effects, possible erosion of soils, and seismic amplification effects, should be conservatively assessed. The design criteria and analyses should be reviewed to ascertain whether the techniques employed are appropriate and represent commonly accepted methods [e.g., U.S. Army Corps of Engineers (1970b)].

The reviewer should examine the spatial variability of the measured properties to ensure that it has been adequately defined. The reviewer should also examine slope characterization data to ensure that nearby slopes, the failure of which could adversely affect the stability of impoundments, have been properly characterized.

The reviewer should determine whether the static and dynamic stability analyses demonstrate that there is an adequate factor of safety against failure.

The reviewer should examine the slope stability analysis to determine that an appropriately conservative approach has been used and that adverse conditions to which the slope might be subjected have been considered. The reviewer should confirm that the static analyses include calculations using appropriate assumptions and methods to assess the following:

(1)   Uncertainties and variations in the shape of the slope, the boundaries and parameters of the several types of soils within the slope, the forces acting on the slope, and the pore pressures acting within and beneath the slope.

(2)   The failure surface corresponding to the lowest factor of safety.

(3)   The effect of the assumptions inherent in the method of analysis used.

The reviewer should ensure that the analysis is conservative and that possible failure modes have been considered, including evaluation of the effect of the maximum credible earthquake, or the appropriate design criteria found acceptable in standard review plan Section 1.4. The reviewer will also verify that the impoundment will not be located near a capable fault on which a maximum credible earthquake larger than that which the impoundment could reasonably be expected to withstand might occur.

The reviewer should be aware that no single method of analysis is applicable for all stability assessments. Therefore, no single method of analysis is recommended. If the staff review indicates that questionable assumptions have been made or that non-standard or inappropriate methods of analysis have been used, the staff may model the slope in a manner consistent with the data and perform an independent analysis.

The reviewer should verify that disposal cell slopes will be relatively flat after final stabilization to minimize the potential for erosion and to provide a conservative factor of safety. In evaluating the slope, the reviewer will focus on determining if the slopes are 5 horizontal to 1 vertical (5h:1v) as required by 10 CFR Part 40, Appendix A, Criterion 4(c). If slopes steeper than 5h:1v are proposed, the reviewer must evaluate these steeper slopes as an alternative to the requirements of Criterion 4(c). In conducting a review of steeper slopes, the reviewer must evaluate the acceptability of the steeper slope using the applicable criteria in this standard review plan and determine if there is an acceptable economic basis and an equivalent level of protection available to justify an alternative to 10 CFR Part 40, Appendix A, Criterion 4(c). The reviewer should evaluate whether a full self-sustaining vegetative cover can be placed over the tailings pile, primarily to reduce the wind and water erosion to negligible levels. If a vegetative cover is not suitable for the site conditions, the reviewer should verify that an appropriate rock cover has been provided. This verification should be coordinated with the review using standard review plan Chapter 3.0.

Because dams at operating facilities, or dams that continue to hold water after the cessation of operations, are also subject to the National Dam Safety Program Act of 1996, the reviewer should determine if the dam is classified as a structure with low hazard potential or high hazard potential. If the dam is classified as high hazard, the reviewer should evaluate the emergency action plan for the facility.

2.2.3 Acceptance Criteria

The analysis of slope stability will be acceptable if it meets the following criteria:

(1)   Slope characteristics are properly evaluated.

    (a)   Cross sections and profiles of natural and cut slopes whose instability would directly or indirectly affect the control of radioactive materials are presented in sufficient number and detail to enable the reviewer to select the cross sections for detailed stability evaluation.

    (b)   Slope steepness is a minimum of five horizontal units (5h) to one vertical unit (1v) or less. The use of slopes steeper than 5h:1v is considered an alternative to the requirements in 10 CFR Part 40, Appendix A, Criterion 4(c). When slopes steeper than 5h:1v are proposed, a technical justification should be offered as to why a 5h:1v or flatter slope would be impractical and compensating factors and conditions are incorporated in the slope design for assuring long-term stability.

    (c)   Locations selected for slope stability analysis are determined considering the location of maximum slope angle, slope height, weak foundation, piezometric level(s), the extent of rock mass fracturing (for an excavated slope in rock), and the potential for local erosion.

(2)   An appropriate design static analysis is presented.

   (a)   The analysis includes calculations with appropriate assumptions and methods of analysis (NRC, 1977). The effect of the assumptions and limitations of the methods used is discussed and accounted for in the analysis. Acceptable methods for slope stability analysis include various limit equilibrium analysis or numerical modeling methods.

   (b)   The uncertainties and variability in the shape of the slope, the boundaries and parameters of the several types of soils and rocks within and beneath the slope, the material properties of soil and rock within and beneath the slope, the forces acting on the slope, and the pore pressures acting within and beneath the slope are considered.

   (c)   Appropriate failure modes during and after construction and the failure surface corresponding to the lowest factor of safety are determined. The analysis takes into account the failure surfaces within the slopes, including through the foundation, if any.

   (d)   Adverse conditions such as high water levels from severe rain and the probable maximum flood are evaluated.

   (e)   The effects of toe erosion, incision at the base of the slope, and other deleterious effects of surface runoff are assessed.

   (f)   The resulting safety factors for slopes analyzed are comparable to the minimum acceptable values of safety factors for slope stability analysis given in NRC Regulatory Guide 3.11 (NRC, 1977).

(3)   Appropriate analyses considering the effect of seismic ground motions on slope stability are presented.

   (a)   Evaluation of overall seismic stability, using pseudostatic analysis or dynamic analysis, as appropriate (U.S. Army Corps of Engineers, 1977; NRC, 1977). Alternatively, a dynamic analysis following Newmark (1965) can be carried out to establish that the permanent deformation of the disposal cell from the design seismic event will not be detrimental to the disposal cell. The reviewer should verify that the yield acceleration or pseudostatic horizontal yield coefficient necessary to reduce the factor of safety against slippage of a potential sliding mass to 1.0 in a "Newmark-type" analysis has been adequately estimated (Seed and Bonaparte, 1992).

   (b)   An appropriate analytical method has been used. A number of different methods of analysis are available (e.g., slip circle method, method of slices, and wedge analysis) with several variants of each (Lambe and Whitman, 1979; U.S. Army Corps of Engineers, 1970b; NRC, 1977; Bromhead, 1992). Limit-equilibrium

2-9

analysis methods do not provide information regarding the variation of strain within the slope and along the slip surface. Consequently, there is no assurance that the peak strength values used in the analysis can be mobilized simultaneously along the entire slip surface unless the material shows ductile behavior (Duncan, 1992). Residual strength values should be evaluated if mobilized shear strength at some points is less than the peak strength. The reviewer should ensure that appropriate conservatism has been incorporated in the analysis using the limit equilibrium methods. The limit equilibrium analysis methodologies may be replaced by other techniques, such as finite element or finite difference methods. If any important interaction effects cannot be included in an analysis, the reviewer must determine that such effects have been treated in an approximate but conservative fashion. The engineering judgment of the reviewer should be used in assessing the adequacy of the resulting safety factors (NRC, 1983a,b).

(c)   For dynamic loads, the dynamic analysis includes calculations with appropriate assumptions and methods (NRC, 1977; Seed, 1967; Lowe, 1967; Department of the Navy, 1982a,b,c; U.S. Army Corps of Engineers, 1970a,b, 1971, 1972; Bureau of Reclamation, 1968). The effect of the assumptions and limitations of the methods used is discussed and accounted for in the analysis.

(d)   For dynamic loads, a pseudostatic analysis is acceptable in lieu of dynamic analysis if the strength parameters used in the analysis are conservative, the materials are not subject to significant loss of strength and development of high pore pressures under dynamic loads, the design seismic coefficient is 0.20 or less, and the resulting minimum factor of safety suggests an adequate margin, as provided in NRC Regulatory Guide 3.11 (NRC, 1977).

(e)   For pseudostatic analysis of slopes subjected to earthquake loads, an assumption is made that the earthquake imparts an additional horizontal force acting in the direction of the potential failure (U.S. Army Corps of Engineers, 1970b, 1977; Goodman, 1989). The critical failure surface obtained in the static analysis is used in this analysis with the added driving force. Minimum acceptable values for safety factors of slope stability analysis are given in Regulatory Guide 3.11 (NRC, 1977).

(f)   The assessment of the dynamic stability considers an appropriate design level seismic event and/or strong ground motion acceleration, consistent with that identified in Chapter 1 of this review plan. Influence of local site conditions on the ground motions associated with the design level event is evaluated. The design seismic coefficient to be used in the pseudostatic analysis is either 67 percent of the peak ground acceleration at the foundation level of the tailings piles for the site or 0.1g, whichever is greater.

(g)   If the design seismic coefficient is greater than 0.20g, then the dynamic stability investigation (Newmark, 1965) should be augmented by other appropriate methods (i.e., finite element method), depending on specific site conditions.

2-10

(h)   In assessing the effects of seismic loads on slope stability, the effect of dynamic stresses of the design earthquake on soil strength parameters is accounted for. As in a static analysis, the parameters such as geometry, soil strength, and hydrodynamic and pore pressure forces are varied in the analysis to show that there is an adequate margin of safety.

(i)   Seismically induced displacement is calculated and documented. There is no universally accepted magnitude of seismically induced displacement for determining acceptable performance of the disposal cell (Seed and Bonaparte, 1992; Goodman and Seed, 1966). Surveys of five major geotechnical consulting firms by Seed and Bonaparte (1992) indicate that the acceptable displacement is from 15 to 30 cm [6 to 12 in.] for tailings piles. The reviewer should ensure that this criterion is also augmented by provisions for periodic maintenance of the slope(s).

(j)   Where there is potential for liquefaction, changes in pore pressure from cyclic loading are considered in the analysis to assess the effect of pore pressure increase on the stress-strain characteristics of the soil and the post-earthquake stability of the slopes. Liquefaction potential is reviewed using Section 2.4 of this review plan. Evaluations of dynamic properties and shear strengths for the tailings, underlying foundation material, radon barrier cover, and base liner system are based on representative materials properties obtained through appropriate field and laboratory tests (NRC, 1978, 1979).

(k)   The applicant has demonstrated that impoundments will not be located near a capable fault on which a maximum credible earthquake larger than that which the impoundment could reasonably be expected to withstand might occur.

(4)   Provision is made to establish a vegetative cover, or other erosion prevention, to include the following considerations:

(a)   The vegetative cover and its primary functions are described in detail.

This determination should be made with respect to any effect the vegetative cover may have on reducing slope erosion and should be coordinated with the reviewer of standard review plan Chapter 3.

If strength enhancement from the vegetative cover is taken into account, the methodology should be appropriate (Wu, 1984).

(b)   In arid and semi-arid regions, where a vegetative cover is deemed not self-sustaining, a rock cover is employed on slopes of the mill tailings. If credit is taken for strength enhancement from rock cover, the reviewer should confirm that appropriate methodology has been presented.

The design of a rock cover, where a self-sustaining vegetative cover is not practical, is based on standard engineering practice. Standard review plan

Geotechnical Stability

Chapter 3 discusses this item in detail.

(5)   Any dams meet the requirements of the dam safety program if the application demonstrates the following:

    (a)   The dam is correctly categorized as a low hazard potential or a high hazard potential structure using the definition of the U.S. Federal Emergency Management Agency.

    (b)   If the dam is ranked as a high hazard potential, an acceptable emergency action plan consistent with the Federal Emergency Management Agency guide (U.S. Federal Emergency Management Agency, 1998) has been developed.

(6)   The use of steeper slopes as an alternative to the requirements in 10 CFR, Part 40, Appendix A, will be found acceptable if the following are met:

    (a)   An equivalent level of stabilization and containment and protection of public health, safety, and the environment is achieved.

    (b)   A site-specific need for the alternate slopes is demonstrated.

2.2.4 Evaluation Findings

If the staff review as described in standard review plan Section 2.2 results in the acceptance of the slope stability, the following conclusions may be presented in the technical evaluation report:

The staff has completed its review of the slope stability at the _____ uranium mill facility. This review included an evaluation using the review procedures in Section 2.2.2 and the acceptance criteria outlined in Section 2.2.3 of this standard review plan.

The licensee has acceptably described the slope stability evaluation by (1) providing cross sections and profiles of natural and cut slopes in sufficient detail and number to represent significant slope and foundation conditions; (2) placing tailings below grade or in demonstrably safe above-grade disposal facilities; (3) ensuring that slope steepnesses are five horizontal (5h) to one vertical (1v) or less or by providing technical justification for a different slope ratio; (4) providing measurements of static and dynamic properties of soil and rock using standards such as those established by the American Society for Testing and Materials, International Society of Rock Mechanics, NRC, or the U.S. Army Corps of Engineers; (5) selecting locations for slope stability analyses while considering the location of maximum slope angle, slope height, weak foundation, the extent of rock mass fracturing, and the potential for local erosion; and (6) describing vegetative cover and its primary functions in detail. Where the licensee has proposed use of steeper slopes as an alternative to the requirements of 10 CFR Part 40, Appendix A, Criterion 4(c), the staff has evaluated the licensee's demonstration that steeper slopes would result in economic savings and also ensure the long-term stabilization of the tailings with a level of protection equivalent to that required in 10 CFR Part 40, Appendix A, Criterion 4(c). Therefore, the use of steeper slopes complies with the alternates requirement in 10 CFR Part 40, Appendix A.

The static loads analysis is acceptable and includes (1) appropriate uncertainties and variabilities in important rock/soils parameters; (2) consideration of appropriate failure modes; (3) a discussion of the effect of the assumptions inherent in the method of analysis used; (4) consideration of adverse conditions, including flooding, with appropriate safety factors; and (5) the effects of toe erosion, incision of the base of the slope, and other deleterious effects of surface runoff.

The dynamic and pseudostatic analyses are acceptable and include (1) calculations with appropriate assumptions and methods; (2) treatment of important interaction effects in a conservative fashion; (3) an accounting of the dynamic stresses of the maximum credible earthquake on soil strength parameters; (4) for pseudostatic analyses of slopes subjected to earthquake loads, consideration of the added driving horizontal force acting in the direction of a potential failure; (5) determination that possible permanent deformation sustained in the slope from a maximum credible earthquake will not damage the effectiveness of the disposal cell; (6) determination that the magnitude of seismically induced displacement does not exceed 15 to 30 cm [6 to 12 in.]; (7) a selection of appropriate design-level seismic events or strong ground motion accelerations; (8) evaluations of local site conditions; (9) evaluations of the potential for liquefaction and the effect of pore pressure increase on the stress-strain characteristics of the soil and post-earthquake stability of the slopes; (10) evaluations of the dynamic properties and shear strength of the tailings, underlying foundation, radon barrier cover, and base liner system; and (11) design of a self-sustaining vegetative or rock cover that is consistent with commonly accepted engineering practice.

On the basis of the information presented in the application and the detailed review conducted of the slope stability at the _____ uranium mill facility, the NRC staff concludes that the slope stability and associated conceptual and numerical models pertaining to design of the impoundments provide an acceptable input to demonstration of compliance with the following criteria in 10 CFR Part 40, Appendix A: Criterion 4(c), which provides requirements for the long-term stability of the embankment and cover slopes for tailings; Criterion 4(d), which requires establishment of a self-sustaining vegetative cover or employment of a rock cover to reduce wind and water erosion to negligible levels, that individual rock fragments are suited for the job, and that the impoundment surfaces are contoured to avoid concentrated surface runoff or abrupt changes in slope gradient; Criterion 4(e), which requires that the impoundment not be located near a capable fault on which a maximum credible earthquake larger than that which the impoundment could reasonably be expected to withstand might occur; Criterion 5(A)(5), which requires the structural integrity of slopes (dikes) to prevent massive failure of the dikes; and Criterion 6(1), which requires that impoundment designs providing reasonable assurance of control of radiological hazards to be effective for 1,000 years to the extent reasonably achievable, and in any case for at least 200 years.

2.2.5 References

American Society for Testing and Materials Standards:

D 2850, "Test Method for Unconsolidated, Undrained Compressive Strength of Cohesive Soils in Triaxial Compression."

D 3080, "Method for Direct Shear Test of Soils Under Consolidated

Drained Conditions."

D 4767, "Test Method for Consolidated-Undrained Triaxial Compression Test on Cohesive Soils."

Bromhead, E.N. *The Stability of Slopes.* London, England: Blackie Academic & Professional. 1992.

Bureau of Reclamation. *Earth Manual.* First Edition. Washington, DC: U.S. Department of the Interior. 1968.

Department of the Navy. "Soil Mechanics." NAVFAC DM 7.1. May 1982a.

————. "Foundations and Earth Structures." NAVFAC DM 7.2. May 1982b.

————. "Soil Dynamics, Deep Stabilization, and Special Geotechnical Construction." NAVFAC DM 7.3. May 1982c.

Duncan, J.M. "State-of-the-Art: Static Stability and Deformation Analysis. Stability and Performance of Slopes and Embankments-II, Volume 1." Proceedings of a Specialty Conference. R.B. Seed and R.W. Boulanger, eds. Geotechnical Special Publication No. 31. New York, New York: American Society of Civil Engineers. 1992.

Goodman, R.E. *Introduction to Rock Mechanics.* 2nd edition. New York, New York: John Wiley and Sons. 1989.

Goodman, R.E. and H.B. Seed. "Earthquake-Induced Displacements in Sand Embankments." *ASCE Journal of the Soil Mechanics and Foundations Division.* Vol. 92, No. SM2. March 1966.

Lambe, T.W. and R.V. Whitman. *Soil Mechanics, SI Version.* New York, New York: John Wiley and Sons. 1979.

Lowe, J., III. "Stability Analysis of Embankments." *ASCE Journal of the Soil Mechanics and Foundations Division.* Vol. 93, No. SM4. 1967.

Newmark, N.M. "Effects of Earthquakes on Dams and Embankments." *Geotechnique.* June 1965.

NRC. NUREG/CR–3199, "Guidance for Disposal of Uranium Mill Tailings: Long-Term Stabilization of Earthen Cover Materials." Washington, DC: NRC. 1983a.

————. NUREG/CR–3397, "Design Considerations for Long-Term Stabilization of Uranium Mill Tailings Impoundments." Washington, DC: NRC. 1983b.

————. Regulatory Guide 1.132, "Site Investigations for Foundations of Nuclear Power Plants." Rev. 1. Washington, DC: NRC, Office of Standards Development. March 1979.

————. Regulatory Guide 1.138, "Laboratory Investigations of Soils for Engineering Analysis and Design of Nuclear Power Plants." Washington, DC: NRC, Office of Standards Development. April 1978.

————. Regulatory Guide 3.11, "Design, Construction, and Inspection of Embankment Retention Systems for Uranium Mills." Rev. 2. Washington, DC: NRC, Office of Standards Development. December 1977.

Seed, H.B. "Slope Stability During Earthquakes." *ASCE Journal of the Soil Mechanics and Foundations Division*. Vol. 93, No. SM4. 1967.

Seed R.B. and R. Bonaparte. "Seismic Analysis and Design of Lined Waste Fills: Current Practice. Stability and Performance of Slopes and Embankments--II, Volume 1." Proceedings of a Specialty Conference. R.B. Seed and R.W. Boulanger, eds. Geotechnical Special Publication No. 31. New York, New York: American Society of Civil Engineers. 1992.

U.S. Army Corps of Engineers. "Earthquake Design and Analysis for Corps of Engineer Dams." ER1110-2-1806. U.S. Department of the Army, Office of the Chief of Engineers. April 1977.

————. "Soil Sampling." Engineering Manual EM1110-2-1907. Department of the Army, Office of the Chief of Engineers. March 1972.

————. "Instrumentation of Earth and Rockfill Dams." Engineering Manual EM1110-2-1908, Part 1 and 2. Department of the Army, Office of the Chief of Engineers. August and November 1971.

————. "Laboratory Soil Testing." Engineering Manual EM1110-2-1906. Department of the Army, Office of Chief Engineers. November 1970a.

————. "Engineering and Design Stability of Earth and Rock Fill Dams." Engineering Manual EM1110-2-1902. Department of the Army, Office of the Chief of Engineers. 1970b.

U.S. Federal Emergency Management Agency. "Federal Guidelines for Dam Safety: Emergency Action Planning for Dam Owners." Washington, DC: U.S. Federal Emergency Management Agency. 1998.

Wu, T.H. "Effect of Vegetation on Slope Stability, "Soil Reinforcement and Moisture Effects on Slope Stability." Transportation Research Record 965. National Research Council, Transportation Research Board. 1984.

2.3 Settlement

2.3.1 Areas of Review

The staff should review the methods and results of testing and analyses conducted to estimate deformation of subsurface materials and uranium mill tailings. This should include examination of material properties and thicknesses of compressible materials, factors used in stress calculations, calculated pore pressures within and beneath the embankment, resulting total and differential settlement of the tailings surface under both static and seismic conditions, and the effects of such settlements on the radon barrier layer of the cover of the disposal cell and erosion protection layer. Liquefaction and associated settlement are addressed in standard review plan Section 2.4. One of the purposes of this review is to determine if the licensee has an acceptable method for determining if tailings consolidation is sufficient to allow the placement of a radon barrier.

2.3.2 Review Procedures

The reviewer should examine the assessments of the magnitudes and distributions of settlement of the disposal cell and the analyses of the potential for cracking of the radon barrier from tensile strains in order to determine the adequacy of the design.

The reviewer should confirm that clay layers and slime in the tailings pile and foundations have been considered in the assessment of both immediate and long-term settlement.

In reviewing the assessment of settlements, the reviewer should give particular attention to the identification and thicknesses of compressible soil layers within the tailings and in the foundation. Settlement should be calculated at several locations within the disposal cell to enable a determination of the overall settlement pattern of the disposal cell cover. The locations for settlement calculations should be selected considering the presence of sand/slime tailings and foundation materials. The tailings are expected to be a hydraulically placed material comprised of interspersed sand and slime tailings. The following specific items should be reviewed to determine the acceptability of the assessment of the magnitudes and distribution of settlement:

(1)   The analysis of immediate settlement of tailings surfaces, considering rebound from excavation and settlement from instantaneous compression of underlying materials and the tailings pile. The computation of incremental tailings loading and the width of the loaded area, as well as the determination of the undrained modulus and Poisson's ratio should be examined. Calculations of the settlement of hydraulically placed tailings should be examined.

(2)   The analysis of consolidation settlement from delayed compression (caused by pore-pressure dissipation) of underlying materials and the tailings pile.

The calculation of settlement should be reviewed to ensure that each compressible soil layer within or underneath the tailings pile is considered and is assigned proper thickness and that the appropriate level of stress change is applied at the mid-depth of the soil layer.

(3)   The estimate of the time at which the primary consolidation settlement of the tailings will be essentially complete. Generally, the radon barrier and disposal cell cover may be placed only after the settlement of tailings is essentially complete.

(4)   The analysis of secondary settlement from long-term creep.

(5)   The distribution of settlement magnitudes for assessment of differential settlement.

(6)   Evaluation of the potential for cracking of the radon barrier layer as a result of long-term settlement of the cover.

2.3.3   Acceptance Criteria

The analysis of tailings settlement will be acceptable if it meets the following criteria:

(1)   Computation of immediate settlement follows the procedure recommended in NAVFAC DM–7.1 (Department of the Navy, 1982). If a different procedure is used, the basis for the procedure is adequately explained.

The procedure recommended in NAVFAC DM–7.1 (Department of the Navy, 1982) for calculation of immediate settlement is adequate if applied incrementally to account for different stages of tailings emplacement. If this method is used, the reviewer should verify that the computation of incremental tailings loading and the width of the loaded area, as well as the determination of the undrained modulus and Poisson's ratio, have been computed and documented.

Settlement of tailings arises from compression of soil layers within the disposal cell and in the underlying materials. Because compression of sands occurs rapidly, compression of sand layers in the disposal cell and foundations must be considered in the assessment of immediate settlement. However, the contribution of immediate settlement to consolidation settlement cannot be ignored. Clay layers and slime undergo instantaneous elastic compression controlled by their undrained stiffness as well as long-term inelastic compression controlled by the processes of consolidation and creep (NRC, 1983a).

(2)   Each of the following is appropriately considered in calculating stress increments for assessment of consolidation settlement:

(a)   Decrease in overburden pressure from excavation

(b)   Increase in overburden pressure from tailings emplacement

Geotechnical Stability

> (c)   Excess pore-pressure generated within the disposal cell
>
> (d)   Changes in ground-water levels from dewatering of the tailings
>
> (e)   Any change in ground-water levels from the reclamation action

(3)   Material properties and thicknesses of compressible soil layers used in stress change and volume change calculations for assessment of consolidation settlement are representative of *in situ* conditions at the site.

(4)   Material properties and thicknesses of embankment zones used in stress change and volume change calculations are consistent with as-built conditions of the disposal cell.

(5)   Values of pore pressure within and beneath the disposal cell used in settlement analyses are consistent with initial and post-construction hydrologic conditions at the site.

(6)   Methods used for settlement analyses are appropriate for the disposal cell and soil conditions at the site. Contributions to settlement by drainage of mill tailings and by consolidation/compression of slimes and sands are considered. Both instantaneous and time-dependent components of total and differential settlements are appropriately considered in the analyses (NRC, 1983a,b,c).

The procedure recommended in NAVFAC DM–7.1 (Department of the Navy, 1982) for calculation of secondary compression is adequate.

(7)   The disposal cell is divided into appropriate zones, depending on the field conditions, for assessment of differential settlement, and appropriate settlement magnitudes are calculated and assigned to each zone.

(8)   Results of settlement analyses are properly documented and are related to assessment of overall behavior of the reclaimed pile.

(9)   An adequate analysis of the potential for development of cracks in the radon/infiltration barrier as a result of differential settlements is provided (Lee and Shen, 1969).

2.3.4 Evaluation Findings

If the staff review, as described in standard review plan Section 2.3, shows that the settlement has no impact on the integrity and functionality of the radon barrier and disposal cell cover, then the following conclusions can be presented in the technical evaluation report. If the settlement impacts the cell cover integrity, then the licensee will be required to revise the design to ensure the functionality of the cell cover before a technical evaluation report can be prepared.

The staff has completed its review of the settlement at the _____ uranium mill facility. This review included an evaluation using the review procedures in Section 2.3.2 and the acceptance criteria outlined in Section 2.3.3 of this standard review plan.

The licensee has acceptably described settlement by presenting computations following the procedure recommended in NAVFAC DM–7.1 (Department of the Navy, 1982) or by explaining the technical merit for an alternate procedure. Material properties, thickness, and load increments used to calculate settlement are representative of site conditions. The applicant has acceptably considered each of the following: (1) decrease in overburden pressure from excavation, (2) increase in overburden pressure from emplaced tailings, (3) excess pore-pressure generated within the tailings disposal cell, (4) changes in ground-water levels from dewatering of the tailings, and (5) changes in ground-water levels from reclamation actions. Pore pressures within and beneath the disposal cell/embankment are consistent with initial and as-built hydrologic site conditions. Methods used to determine settlement are appropriate for the tailings embankment and soil conditions at the site. The results of the settlement analyses are properly documented. The tailings embankment has been subdivided acceptably into assessment zones with appropriately assigned settlement magnitudes. The settlement data provide information to assess the possibility of surface ponding or sudden change of gradient caused by settlement. An acceptable analysis for the development of cracks in the radon/infiltration barrier is provided.

On the basis of information presented in the application and the detailed review conducted of the characteristics of the settlement at the _____ mill facility, the NRC staff concludes that the settlement and associated conceptual and numerical models present information needed to demonstrate compliance with 10 CFR Part 40, Appendix Ae. Criterion 6(1), which requires that impoundment designs provide reasonable assurance of control of radiological hazards to be effective for 1,000 years to the extent reasonably achievable, and in any case for at least 200 years.

2.3.5 References

American Society for Testing and Materials Standards:

> D 2435, "Test Method for One-Dimensional Consolidation Properties of Soil."

> D 4719, "Test Method for Pressuremeter Testing in Soils."

Department of the Navy. 1982. *Soil Mechanics.* NAVFAC DM–7.1. May 1982.

Lee, K.L. and C.K. Shen. "Horizontal Movements Related to Subsidence." *ASCE Journal of Soil Mechanics and Foundations Division.* Vol. 95, No. SM1. 1969.

NRC. 1983a. NUREG/CR–3204, "Consolidation of Tailings." Washington, DC: NRC. 1983a.

———. NUREG/CR–3199, "Guidance for Disposal of Uranium Mill Tailings: Long-Term Stabilization of Earthen Cover Materials." Washington, DC: NRC. 1983b.

———. NUREG/CR–3397, "Design Considerations for Long-Term Stabilization of Uranium Mill Tailings Impoundments." Washington, DC: NRC. 1983c.

2.4   Liquefaction Potential

2.4.1   Areas of Review

The staff should review the analysis of the liquefaction potential of subsurface, pile, and embankment materials, and the associated test and data interpretations. Consequences of the liquefaction of subsurface soils and/or uranium mill tailings affecting the settlements within and stability of the disposal cell and the erosion protection layer should also be reviewed. Design features or mitigation actions that address liquefaction potential should be examined. The effect of settlements not induced by liquefaction is considered in standard review plan Section 2.3 and is also considered in standard review plan Section 2.4.3.

2.4.2   Review Procedures

The reviewer should examine the analysis of liquefaction potential by studying the results of geotechnical investigations and *in situ* tests such as standard penetration, cone penetration, piezocone, density, and strength tests as well as boring logs, laboratory classification test data, water table measurements, perched water zones, and soil profiles, to determine if any of the site soils or the tailings pile material could be susceptible to liquefaction.

If it is determined that there may be soils susceptible to liquefaction beneath the site or in the tailings pile, the reviewer should examine the adequacy of site exploration programs, the laboratory test program, and the analyses. Where global liquefaction potential exists, the reviewer should determine that it has been mitigated or eliminated. Minor or local liquefaction potential should be accounted for in settlement analyses.

The reviewer should compare the liquefaction potential analysis in the reclamation plan to an independent study performed by the staff, if necessary.

2.4.3   Acceptance Criteria

The analysis of the liquefaction potential will be acceptable if the following criteria are met:

(1)   Applicable laboratory and/or field tests are properly conducted (NRC, 1978, 1979; U.S. Army Corps of Engineers, 1970, 1972).

(2)   Data for all relevant parameters for assessing liquefaction potential are adequately collected and the variability has been quantified.

(3)   Methods used for interpretation of test data and assessment of liquefaction potential are consistent with current practice in the geotechnical engineering profession (Seed and Idriss, 1971, 1982; National Center for Earthquake Engineering Research, 1997). An assessment of the potential adverse effects that complete or partial liquefaction could have on the stability of the embankment may be based on cyclic triaxial test data obtained from undisturbed soil samples taken from the critical zones in the site area (Seed and Harder, 1990; Shannon & Wilson, Inc. and Agbabian-Jacobsen

2-20

Associates, 1972).

(4)   If procedures based on laboratory tests combined with ground response analyses are used, laboratory test results are corrected to account for the difference between laboratory and field conditions (NRC, 1978; Naval Facility Engineering Command, 1983).

(5)   The time history of earthquake ground motions used in the analysis is consistent with the design seismic event.

(6)   If the potential for complete or partial liquefaction exists, the effects such liquefaction could have on the stability of slopes and settlement of tailings are adequately quantified.

(7)   If a potential for global liquefaction is identified, mitigation measures consistent with current engineering practice or redesign of tailings ponds/embankments are proposed and the proposed measures provide reasonable assurance that the liquefaction potential has been eliminated or mitigated.

(8)   If minor liquefaction potential is identified and is evaluated to have only a localized effect that may not directly alter the stability of embankments, the effect of liquefaction is adequately accounted for in analyses of both differential and total settlement and is shown not to compromise the intended performance of the radon barrier. Additionally, the disposal cell is shown to be capable of withstanding the liquefaction potential associated with the expected maximum ground acceleration from earthquakes. The licensee may use post-earthquake stability methods (e.g., Ishihara and Yoshimine, 1990) based on residual strengths and deformation analysis to examine the effects of liquefaction potential. Furthermore, the effect of potential localized lateral displacement from liquefaction, if any, is adequately analyzed with respect to slope stability and disposal cell integrity.

2.4.4   Evaluation Findings

If the staff review, as described in standard review plan Section 2.4, results in the acceptance of the licensee liquefaction potential analysis and conclusions on the impact on the performance of the disposal cell, the following conclusions may be presented in the technical evaluation report:

The staff has completed its review of the liquefaction potential at the _____ uranium mill facility. This review included an evaluation using the review procedures in standard review plan Section 2.4.2 and acceptance criteria outlined in Section 2.4.3 of this standard review plan.

The licensee has acceptably evaluated liquefaction potential based on results from properly conducted laboratory and/or field tests. The methods used for interpretation of test data are consistent with current practice. Where global liquefaction is identified, mitigation measures or redesign of tailings ponds/embankments are proposed and the new design provides reasonable assurance that the liquefaction potential has been eliminated or mitigated. In the case of minor/local liquefaction potential, its effect is accounted for in the analysis of both differential

Geotechnical Stability

and total settlement and is shown not to compromise the intended performance of the radon barrier and erosion protection.

On the basis of the information presented in the application and the detailed review conducted of the liquefication potential at the _____ uranium mill facility, the NRC staff concludes that the results of evaluation of liquefaction potential and associated conceptual and numerical models present input to a demonstration of compliance with the following criteria in 10 CFR Part 40, Appendix A: Criterion 4(c), which provides long-term stability requirements for the slopes of the tailings embankment and cover; and Criterion 6(1), which requires that impoundment designs povide reasonable assurance of control of radiological hazards to be effective for 1,000 years to the extent reasonably achievable, and in any case for at least 200 years.

2.4.5 References

American Society for Testing and Materials Standards:

> D 3999, "Test Method for the Determination of the Modulus and Damping Properties of Soils Using the Cyclic Triaxial Apparatus."

> D 4015, "Test Method for Modulus and Damping of Soils by the Resonant-Column Method."

Ishihara, K. and M. Yoshimine. "Evaluation of Settlements in Sand Deposits Following Liquefaction During Earthquake." *Soil Foundations*. Vol. 32, No. 1. Japanese Society of Soil Mechanics and Foundation Engineering. 1990.

National Center for Earthquake Engineering Research. "Proceedings of the NCEER Workshop on Evaluation of Liquefaction Resistance of Soils." T.L. Youd and I.M. Idriss, eds. Technical Report No. NCEER 97-002. Buffalo, New York: State University of New York. 1997.

Naval Facility Engineering Command. "Soil Dynamics, Deep Stabilization, and Special Geotechnical Construction. NAVFAC DM–7.3. Alexandria, Virginia: Department of the Navy. 1983.

NRC. Regulatory Guide 1.132, "Site Investigations for Foundations of Nuclear Power Plants." Rev. 1. Washington, DC: NRC, Office of Standards Development. March 1979.

———. Regulatory Guide 1.138, "Laboratory Investigations of Soils for Engineering Analysis and Design of Nuclear Power Plants." Washington, DC: NRC, Office of Standards Development. April 1978.

Seed, H.B. and I.M. Idriss. "Ground Motions and Soil Liquefaction During Earthquakes." Earthquake Engineering Research Institute. *Engineering Monograph*: 5. 1982.

Seed, H.B. and I.M. Idriss. "A Simplified Procedure for Evaluating Soil Liquefaction Potential." *Journal of Soil Mechanics and Foundation Division*. Vol. 97, No. SM 9. pp. 1,249–1,274. 1971.

Seed, R.B. and L.F. Harder. "SPT-Based Analysis of Cyclic Pore Pressure Generation and Undrained Residual Strength." *Proceedings of the H. Bolton Seed Memorial Symposium.* Berkeley, California: University of California, May 10–11. pp. 351–376. 1990.

Shannon & Wilson, Inc. and Agbabian-Jacobsen Associates. "Soil Behavior Under Earthquake Loading Conditions: State-of-the-Art Evaluation of Characteristics for Seismic Responses Analyses." Washington, DC: U.S. Atomic Energy Commission. 1972.

U.S. Army Corps of Engineers. "Soil Sampling." Manual EM 1110-2-1907. March 1972.

———. "Laboratory Soil Testing, Engineering." Manual EM1110-2-1906. November 1970.

2.5   Disposal Cell Cover Engineering Design

2.5.1   Areas of Review

The staff should review information presented on disposal cell cover engineering design. including field exploration data, laboratory test results, design details, and construction and installation considerations pertinent to the geotechnical aspects of design and any associated geomembranes (i.e., disposal cell configuration and thickness, compaction requirements, gradations, permeability, and dispersivity).

2.5.2   Review Procedures

The reviewer should examine the disposal cell design and engineering parameters to assess the geotechnical aspects of the disposal cell cover. Specific aspects of the review should consider the following items:

(1)   Determination that an adequate quantity of the specified borrow material has been identified at the borrow source.

(2)   Confirmation that placement density, specific gravity, moisture content, dispersivity, and shrinkage properties used in the disposal cell design have been determined by suitable laboratory testing so that long-term stability standards will be met. (Note that permeability issues are discussed separately in standard review plan Section 2.7.)

(3)   Confirmation that appropriate measures for controlling the effects of erosion, surface water flows, and vegetative deep root penetrations have been taken.

(4)   Verification that the particle size gradation of the disposal cell cover material, bedding layers, other layers in the cover, and the rock layer are compatible to ensure stability against particle migration during the period of regulatory interest.

(5)   Determination that the disposal cell has been designed to accommodate the effects of anticipated freeze-thaw cycles.

(6)   Assessment, if bentonite amendment to the radon barrier material of the disposal cell cover is proposed, of whether supporting discussions define appropriate laboratory

Geotechnical Stability

testing and field procedures associated with evaluating amended materials.

(7)   Determination if the cracking potential of the disposal cell has been adequately addressed. Cracking from both settlement and shrinkage should be evaluated using standard review plan Section 2.3.

(8)   Assessment of the acceptability of plans for installation and use of any geomembranes.

(9)   Confirmation that the information used in the disposal cell cover design appropriately reflects the staff findings on the information reviewed using standard review plan Chapters 1.0, 2.0, 3.0, and 4.0.

Note that hydraulic conductivity aspects of the disposal cell cover design are assessed using standard review plan Section 2.7 and that review of the disposal cell design features is addressed in standard review plan Sections 2.2, 2.3, and 2.4. Review of the radon attenuation aspects of the disposal cell design is addressed in standard review plan Chapter 5.0.

2.5.3   Acceptance Criteria

The assessment of the disposal cell cover design and engineering parameters will be acceptable if it meets the following criteria:

(1)   Detailed descriptions of the disposal cell material types [e.g., Unified Soil Classification System (Holtz and Kovacs, 1981)] and/or soil mixtures (e.g., bentonite additive) and the basis for their selection are presented.

An analysis is included demonstrating that an adequate quantity of the specified borrow material has been identified at the borrow source. The information on borrow material includes boring and test pit logs and compaction test data.

The soils that are considered suitable include the Unified Classification System Classes CL, CH, SC, and CL-ML, with desirable characteristics and limitations as listed in Table 3-1 of the "Construction Methods and Guidance for Sealing Penetrations in Soil Covers" (Bennett and Homz, 1991; Bennett and Kimbrell, 1991). The preferred material for the low-permeability layers is inorganic clay soil. This soil should be compacted to a low saturated hydraulic conductivity of at least 1×10^{-7} cm/sec. For drainage layers, cobble types GW, GP, SP, and SW are recommended, with GW and GP being the preferred types (Bennett, 1991).

Measures for resisting cracking, heaving, and settlement, and providing protection from burrowing animals, root penetration, and erosion over a long period of time are described.

(2)   A sufficiently detailed description of the applicable field and laboratory investigations and testing that were completed, and the material properties (e.g., permeability, moisture-density relationships, gradation, shrinkage and dispersive characteristics, resistance to freeze-thaw degradation, cracking potential, and chemical compatibility, including any amendment materials) are identified (U.S. Army Corps of Engineers,

1970, 1972; Fermulk and Haug, 1990; NRC, 1978, 1979; Lee and Shen, 1969; Spangler and Handy, 1982).

(3)   Details are presented (including sketches) of the disposal cell cover termination at boundaries, with any considerations for safely accommodating subsurface water flows.

(4)   A schematic diagram displaying various disposal cell layers and thicknesses is provided.

The particle size gradation of the disposal cell bedding layer and the rock layer are established to ensure stability against particle migration during the period of regulatory interest (NRC, 1982).

(5)   The effect of possible freeze-and-thaw cycles on soil strength and radon barrier effectiveness is adequately considered (e.g., Aitken and Berg, 1968).

If the region experiences prolonged freezing, the disposal cell cover may be affected by the freeze-thaw cycle. During freezing, ice crystals and lenses can form in the soil, causing heaving. On the other hand, during melting and thawing, the soil may lose its bearing capacity because of development of supersaturated conditions (Spangler and Handy, 1982). Major factors affecting growth of ice in soil are the temperature below the freezing point, the capillary characteristics of the soil, and the presence of water. The reviewer should check whether the soil is susceptible to frost heave, considering that uniformly graded soils containing more than 10 percent of particles smaller than 0.02 mm and well-graded soils with more than 3 percent of particles smaller than 0.02 mm are susceptible (Holtz and Kovacs, 1981; Spangler and Handy, 1982). After many freeze-thaw cycles, the soil may become a loose collection of aggregates with significantly reduced overall strength.

(6)   A description is given (with sketches) of any penetrations (e.g., monitoring wells) through the disposal cell system, including details of penetration sealing and disposal cell cover integrity. Bennett and Kimbrell (1991) suggest methods for seal design that are acceptable.

(7)   An adequate analysis is presented of the potential for development of cracks in the disposal cell cover as a result of differential settlement and shrinkage. Note that cracking issues associated with settlement are discussed in standard review plan Section 2.3.3.

(8)   An adequate description of the geomembranes and their major properties (e.g., physical, mechanical, and chemical) is provided if low permeability geomembranes are proposed as a part of the disposal cell cover. Methods for installation of the membranes in accordance with the manufacturer's recommendations are discussed. The shear strength of the interface between compacted clay and geomembranes used in the stability analyses under both static and dynamic loads is noted. The expected service life of the geomembrane is analyzed.

(9)   Information on site characterization, slope stability, settlement, and liquefaction used in the disposal cell cover design appropriately reflects the Licensee's evaluation, and

Geotechnical Stability

> therefore, constitutes inputs that would contribute to the demonstration of disposal cell design compliance with the regulations.

2.5.4 Evaluation Findings

If the staff review as described in standard review plan Section 2.5 results in the acceptance of the disposal cell cover design, the following conclusions may be presented in the technical evaluation report:

The staff has completed its review of the disposal cell cover design at the _____ uranium mill facility. This review included an evaluation using the review procedures in Section 2.5.2 and acceptance criteria outlined in Section 2.5.3 of this standard review plan.

The licensee has acceptably defined the disposal cell cover design by presenting detailed descriptions of the disposal cell material types and/or soil mixtures, including the basis for their selection. The applicant has identified an adequate quantity of the specified borrow material at the borrow source. An acceptable schematic diagram displaying various disposal cell layers and thicknesses is provided. A description of the applicable field and laboratory investigations and testing is provided, including identification of material properties. The properties of the cover materials have been measured properly using standards such as American Society for Testing and Materials, NRC, or U.S. Army Corps of Engineers. Details (including sketches) have been provided of (1) disposal cell termination boundaries; (2) penetrations, including sealing and disposal cell integrity; and (3) geomembranes and their physical, mechanical, and chemical properties. Methods of installation for the membranes have been discussed and the expected service life has been justified. The analysis of the potential for development of cracks in the disposal cell cover is acceptable.

On the basis of the information presented in the application and the detailed review conducted of the disposal cell cover design at the _____ uranium mill facility, the NRC staff concludes that the disposal cell engineering parameters and associated conceptual and numerical models are acceptable and provide input to demonstration of compliance with the following criteria in 10 CFR, Part 40, Appendix A: Criterion 4(c), which provides requirements for the embankment and cover slopes for tailings; and Criterion 6(1), which requires that impoundment design provide reasonable assurance of control of radiological hazards to be effective for 1,000 years to the extent reasonably achievable, and in any case, for at least 200 years.

2.5.5 References

American Society for Testing and Materials Standards:

> D 75, "Practice for Sampling Aggregates."

> D 4992, "Practice for the Evaluation of Rock To Be Used for Erosion Control."

Aitken, G.W. and R.L. Berg. "Digital Solution of Modified Berggren Equation to Calculate Depths of Freeze or Thaw in Multi-layered Systems." Special Report 122. Hanover, New

Hampshire: Cold Regions Research & Engineering Laboratory. 1968.

Bennett, R.D. NUREG/CR–5432, "Recommendations to the NRC for Soil Cover Systems Over Uranium Mill Tailings and Low-Level Radioactive Wastes: Identification and Ranking of Soils for Disposal Facility Covers." Vol. 1. Washington, DC: NRC. 1991.

Bennett, R.D. and R.C. Homz. NUREG/CR–5432, "Recommendations to the NRC for Soil Cover Systems Over Uranium Mill Tailings and Low-Level Radioactive Wastes: Laboratory and Field Tests for Soil Covers." Vol. 2. Washington, DC: NRC. 1991.

Bennett, R.D. and A.F. Kimbrell. NUREG/CR–5432, "Recommendations to the NRC for Soil Cover Systems Over Uranium Mill Tailings and Low-Level Radioactive Wastes: Construction Methods for Sealing Penetrations in Soil Covers." Vol. 3. Washington, DC: NRC. 1991.

Fermulk, N. and M. Haug. "Evaluation of *In Situ* Permeability Testing Methods." *ASCE Journal of Geotechnical Engineering*. Vol. 116, No. 2. pp. 297–311. 1990.

Holtz, R.D. and W.D. Kovacs. *An Introduction to Geotechnical Engineering*. Englewood Cliffs, New Jersey: Prentice-Hall. 1981.

Lee, K.L. and C.K. Shen. 1969. "Horizontal Movements Related to Subsidence." *Journal of Soil Mechanics and Foundation Division*. Vol. 95, No. SM–1. New York, New York: American Society of Civil Engineers. 1969.

NRC. NUREG/CR–2684, "Rock Riprap Design Methods and Their Applicability to Long-Term Protection of Uranium Mill Tailings Impoundments." Washington, DC: NRC. 1982.

––––––. Regulatory Guide 1.132, "Site Investigations for Foundations of Nuclear Power Plants." Rev. 1. Washington, DC: NRC, Office of Standards Development. March 1979.

––––––. Regulatory Guide 1.138, "Laboratory Investigations of Soils for Engineering Analysis and Design of Nuclear Power Plants." April 1978. Washington, DC: NRC, Office of Standards Development. April 1978.

Spangler, M.G. and R.L. Handy. *Soil Engineering*. New York, New York: Harper and Row. 1982.

U.S. Army Corps of Engineers. "Soil Sampling." Engineering Manual EM1110-2-1907. March 1972.

––––––. "Laboratory Soil Testing." Engineering Manual EM1110–2–1906. November 1970.

2.6   Construction Considerations

2.6.1   Areas of Review

The staff should review information on the geotechnical aspects of reclamation construction. These aspects should include details such as the sequence and schedule for construction activities, material specifications and placement procedures, and quality control aspects of the construction procedures. The geotechnical aspects of the planned construction operations should be reviewed to identify any deviations from standard engineering practice for earthworks, including measures to protect against erosion and provisions for a vegetative cover, if appropriate.

2.6.2 Review Procedures

The reviewer should determine if all the tailings and contaminated materials at the site can be placed within the configuration of the proposed stabilized pile. The construction sequence should be reviewed to verify the feasibility of achieving the intended final configuration of the tailings, particularly when tailings are to be relocated to new areas of the remediated pile, and to determine whether the schedule for completion is reasonable. The reviewer should also confirm that the construction schedule will allow the radon barrier to be completed as expeditiously as practical after ceasing operations.

The reviewer should examine material placement, placement moisture content (drying, if needed), placement density, and desired permeability to ensure that design specifications will be met. If mixing of the fine tailings (slimes) with sand tailings is proposed, the specifications to control the mixture and the determination of the engineering properties of this mixture should be examined for adequacy.

The reviewer should examine the proposed construction quality control program to verify that adequate provisions have been included to ensure that the construction will be in accordance with the NRC-approved reclamation plan. In particular, details of the proposed testing and inspection program, including the type and frequency of tests proposed, should be reviewed and compared with NRC guidance on testing and inspection.

Methods and schedules for emplacing the vegetative cover should be reviewed if necessary, to determine that they are reasonable, and that seeds for the planned vegetation are compatible with the local climate.

2.6.3 Acceptance Criteria

The analysis of construction considerations will be acceptable if the following criteria are met:

(1)   Engineering drawings are at appropriate scales to completely and clearly show the design features (e.g., embankments, riprap, and channels).

(2)   Sources and quantities of borrow material are identified, are shown to have been adequately characterized and quantified through field and laboratory tests, and are demonstrated to be adequate for meeting the geotechnical design requirements for the disposal cell (NRC, 1978, 1979). The background levels of contamination in the borrow materials, if any, are properly established.

(3)   Methods, procedures, and requirements for excavating, hauling, stockpiling, and placing of contaminated and non-contaminated materials and other disposal cell materials are

provided and are shown to be consistent with commonly accepted engineering practice for earthen works (Department of the Navy, 1982a,b; Denson, et al., 1987).

Material placement and compaction procedures are adequate to achieve the desired moisture content (drying, if needed) placement density and permeability. Recommendations made in NUREG/CR–5041 (Denson, et al., 1987) for gradation, placement, and compaction necessary to achieve design drainage rates and volumes, prevent internal erosion or piping, and allow for collection and removal of liquids, are acceptable. Compaction specifications include restrictions on work related to adverse weather conditions (e.g., rainfall, freezing conditions).

Specifications for controlling the mixture of fine tailings (slime) with sand tailings are consistent with commonly accepted engineering practice and testing programs for determination of engineering properties of this mixture.

(4)   A plan for embankment construction is presented, that demonstrates embankments can be constructed in accordance with the design.

(5)   Plans, specifications, and requirements for disposal cell compaction are supported by field and laboratory tests and analyses to assure stability and reliable performance.

(6)   Testing and surveying programs to determine the extent of cleanup required are adequate. The contamination cleanup plan includes the method for determining the extent of the contaminated area and a confirmation program to demonstrate that the contaminated material has been removed. Details of the site cleanup (radiological aspects) are addressed in standard review plan Chapter 5.0.

(7)   A plan for settlement measurement is provided that is satisfactory for producing representative settlement data throughout the area of the disposal cell. Settlement measurement stations are of sufficient coverage and are strategically placed to yield adequate information for determination of total, differential, and residual settlements. Monitoring monuments are designed to be durable. The reviewer should also determine the reasonableness of the proposed monitoring frequency in accordance with NUREG/CR–3356 (NRC, 1983). In the past, the staff has determined that the final radon barrier may be emplaced once 90 percent of expected settlement has occurred.

(8)   All tailings and contaminated materials at the site can be placed within the planned configuration of the stabilized pile.

(9)   Procedures, specifications, and requirements for riprap, rock mulch, and filter production and placement are provided and are shown to be consistent with commonly accepted engineering practice and the design specifications (NRC, 1977, 1982).

(10)   The construction sequence is described and demonstrated to be adequate to achieve the intended configuration for the tailings, particularly when tailings are to be relocated to new areas of the reclaimed pile. The proposed time to completion has been shown to be reasonably achievable, and the construction schedule provides for completing the

radon barrier as expeditiously as practical after ceasing operations in accordance with an approved reclamation plan.

(11)   The vegetation program or rock cover design is described and demonstrated to be adequate (Wu, 1984; NRC, 1982).

(12)   Appropriate quality control provisions are provided to ensure that the construction will be in accordance with the reclamation plan. The descriptions of the methods, procedures, and frequencies by which the construction materials and activities are to be tested and inspected are reasonable and appropriate records will be maintained (NRC, 1983).

(13)   Tailings are placed below grade, or the licensee has demonstrated that the above-grade disposal design provides reasonably equivalent isolation of the tailings from natural erosional forces. Tailings pile topographic features take into account wind protection and vegetation cover.

2.6.4 Evaluation Findings

If the staff review as described in this section results in the acceptance of the licensee proposed construction considerations, the following conclusions may be presented in the technical evaluation report.

The staff has completed its review of construction considerations at the _____ uranium mill facility. This review included an evaluation using the review procedures in Section 2.6.2 and the acceptance criteria outlined in Section 2.6.3 of this standard review plan.

The licensee has acceptably described the construction considerations by (1) providing complete engineering drawings showing all design features; (2) describing sources and quantities of borrow material, including acceptable field and laboratory testing; and (3) identifying methods, procedures, and requirements for excavations, haulage, stockpiling, and placement of materials and demonstrating that all are consistent with accepted engineering practices for earthen works. An acceptable plan for embankment construction is provided. Disposal cell compaction plans are supported by field and laboratory tests that assure stability and performance. The licensee has an acceptable program to determine the extent of cleanup using appropriate testing and surveying programs. An acceptable plan for settlement measurement is provided, including (1) proper coverage and placement of settlement measurement stations, (2) durable monitoring monuments, and (3) reasonable monitoring frequencies. All tailings and contaminated materials have been demonstrated to fit within the planned configuration of the stabilized pile. Procedures, specifications, and requirements for riprap, rock mulch, and filters are provided and are shown to be consistent with commonly accepted engineering practices and design specifications. An acceptable construction sequence, including a reasonable time to completion, has been described. An acceptable vegetation program or rock cover design is proposed. Appropriate quality control provisions are in place to ensure that construction will be in accordance with the reclamation plan and that appropriate records will be maintained.

On the basis of the information presented in the application and the detailed review conducted of the construction considerations at the _____ uranium mill facility, the NRC staff concludes that the construction considerations and associated conceptual and numerical models provide input to a demonstration of compliance with the following criteria in 10 CFR, Part 40, Appendix A: Criterion 4(c), which provides requirements for the embankment and cover slopes for tailings; Criterion 4(d), which requires establishment of a self-sustaining vegetative cover or employment of a rock cover to reduce wind and water erosion to negligible levels, that individual rock fragments are suited for the job, and that the impoundment surfaces are contoured to avoid concentrated surface runoff or abrupt changes in slope gradient; Criterion 6(1), which requires that impoundment designs provide reasonable assurance of control of radiological hazards to be effective for 1,000 years to the extent reasonably achievable, and in any case, for at least 200 years; and Criterion 6A(1), which requires that the radon barrier be completed as expeditiously as practical after ceasing operations in accordance with a Commission-approved reclamation plan.

2.6.5 References

American Society for Testing and Materials Standards:

D 698, "Test Method for Laboratory Compaction Characteristics of Soil Using Standard Effort."

D 1556, "Test Method for Density and Unit Weight of Soil In Place by the Sand Cone Method."

D 1557, "Test Method for Laboratory Compaction Characteristics of Soil Using Modified Effort."

D 2167, "Test Method for Density and Unit Weight of Soil In Place by the Rubber Balloon Method."

D 2922, "Test Methods for Density of Soil and Soil-Aggregate in Place by Nuclear Methods (Shouldow Depth)."

D 2937, "Test Method for Density of Soil in Place by the Drive Cylinder Method."

D 3017, "Test Method for Water Content of Soil and Rock in Place by Nuclear Methods (Shouldow Depth)."

D 3740, "Practice for the Evaluation of Agencies Engaged in the Testing and/or Inspection of Soil and Rock as Used in Engineering Design and Construction."

D 4253, "Test Methods for Maximum Index Density and Unit Weight of Soils Using a Vibratory Table."

D 4254, "Test Method for Minimum Index Density and Unit Weight of Soils and Calculation of Relative Density."

Geotechnical Stability

D 4643, "Test Method for Determination of Water (Moisture) Content of Soil by the Microwave Oven Method"

D 4718, "Practice for Correction of Unit Weight and Water Content for Soils Containing Oversize Particles."

D 4914, "Test Methods for Density of Soil and Rock In Place by the Sand Replacement Method in a Test Pit."

D 5030, "Test Method for Density of Soil and Rock in Place by the Water Replacement Method in a Test Pit."

Denson, R.H., et al. NUREG/CR–5041, "Recommendations to the NRC for Review Criteria for Alternative Methods of Low-Level Radioactive Waste Disposal." Washington, DC: NRC. 1987.

Department of the Navy. "Foundations and Earth Structures." NAVFAC DM–7.2. May 1982a.

———. "Soil Dynamics, Deep Stabilization, and Special Geotechnical Construction." NAVFAC DM–7.3. May 1982b.

NRC. NUREG/CR–3356, "Geotechnical Quality Control: Low-Level Radioactive Waste and Uranium Mill Tailings Disposal Facilities." Washington, DC: NRC. 1983.

———. NUREG/CR–2684, "Rock Riprap Design Methods and Their Applicability to Long-Term Protection of Uranium Mill Tailings Impoundments." Washington, DC: NRC. 1982.

———. Regulatory Guide 1.132, "Site Investigations for Foundations of Nuclear Power Plants." Rev. 1. Washington, DC: NRC, Office of Standards Development. March 1979.

———. Regulatory Guide 1.138, "Laboratory Investigations of Soils for Engineering Analysis and Design of Nuclear Power Plants." Washington, DC: NRC, Office of Standards Development. April 1978.

———. Regulatory Guide 3.11, "Design, Construction, and Inspection of Embankment Retention Systems for Uranium Mills." Rev. 2. Washington, DC: NRC, Office of Standards Development. 1977.

Wu, T.H. "Effect of Vegetation on Slope Stability: Soil Reinforcement and Moisture Effects on Slope Stability." Transportation Research Record 965. National Research Council, Transportation Research Board. 1984.

2.7   Disposal Cell Hydraulic Conductivity

2.7.1 Areas of Review

The staff should review test results, calculations, the technical bases for disposal cell design hydraulic conductivity values, the field testing program, and the quality control program.

2.7.2 Review Procedures

The reviewer should examine the geotechnical design aspects of the disposal cell to ensure that the disposal cell cover component has a minimal hydraulic conductivity, to limit radon emissions from, and water infiltration into, stabilized mill tailings. The geotechnical reviewer should coordinate with the water resources protection reviewer (see standard review plan Chapter 4.0) to ensure that regulatory requirements for ground-water protection can be met by the proposed radon barrier.

The reviewer should verify that an adequate technical basis has been presented for the design hydraulic conductivity (K) value for the disposal cell cover. For any situation in which a $K<10^{-7}$ cm/sec is proposed by the licensee, the staff should verify that either a test fill program will be undertaken to verify the constructability to achieve the desired K value, or the reclamation plan narrative and accompanying analyses have adequately demonstrated the acceptability of the design K value, considering technical papers on this subject (e.g., Rogowski, 1990; Panno, et al., 1991; Benson and Daniel, 1990). If the reclamation plan acceptably demonstrates that field testing is not required, the reviewer should document the technical basis in the technical evaluation report. If field testing is required, the staff should ensure that the test fill specifications require that the hydraulic conductivity value be verified by in-place testing with double-ring infiltrometers or other approved methods.

The test reviewer should examine the test fill construction plan and verification program for adequacy, including such aspects as (1) use of proper procedures and equipment for placement and compaction operations; (2) verification of the material and thickness for the barrier test zone; (3) comparison of gradation, bentonite amendment, and moisture/density testing with specifications; (4) review of the quality control plan; and (5) review of the proposed construction schedule.

2.7.3 Acceptance Criteria

The analysis of disposal cell hydraulic conductivity will be acceptable if it meets the following criteria:

(1)   A sufficient technical basis is provided for the design hydraulic conductivity (K) value for the disposal cell.

The hydraulic conductivity is minimized by compacting fine-grained soil for a sufficient depth above the stabilized tailings. Natural borrow soils having insufficient silt and clay content to effectively reduce the hydraulic conductivity of the barrier can be amended with bentonite for improved effectiveness. (Note that construction issues are discussed

2-33

Geotechnical Stability

separately using standard review plan Section 2.6.)

(2)   A field testing program adequate to verify the constructability of the disposal cell with a design hydraulic conductivity K<10^{-7} cm/sec is provided unless the reclamation plan demonstrates that field testing is not required (Benson and Daniel, 1990; NRC, 1979).

To meet the U.S. Environmental Protection Agency (EPA) ground-water standards, designers of disposal cells for mill tailings sites are proposing increasingly smaller design hydraulic conductivity (K) values. It is not unusual for laboratory permeability test values to yield results of 10^{-8} to 10^{-10} cm/sec. Such tests are performed on compacted soil samples considered by the design engineer to represent the soil to be used for the disposal cell. However, several technical papers (Rogowski, 1990; Panno et al., 1991; Benson and Daniel, 1990) have raised serious questions concerning the exclusive use of laboratory testing for demonstrating hydraulic conductivity values in those cases in which a radon barrier K-value less than 10^{-7} cm/sec is specified. On the basis of these technical papers, field testing is necessary to confirm the radon barrier hydraulic conductivity, since construction operations and soil material variability can create preferred pathways, joints, seams, holes, and flaws that effectively increase the value of this parameter. Test results should take into consideration the variability and uncertainty in site conditions and material properties. The test results should be properly documented and available for inspection.

(3)   An appropriate quality control program is followed for the field testing to determine hydraulic conductivity (NRC, 1983).

For all cases in which K<10^{-7} cm/sec and the test fill program requirement has been defined, specifications and related documents (Remedial Action Inspection Plan, etc.) will require an adequate quality control program. An acceptable quality control program should contain mechanisms to ensure that as-built construction duplicates the test fill construction techniques on the cell barrier (NRC, 1983). The objective of the quality control program will be to provide assurance that uniform and high-quality construction of the cell barrier has been achieved. Records for implementation of the quality control program during the construction of the cell barrier should be properly maintained and available for inspection.

(4)   A reasonable construction schedule is proposed. The proposed construction schedule should promote completion of the radon barrier as expeditiously as practical after ceasing operations in accordance with a written, Commission-approved reclamation plan.

2.7.4 Evaluation Findings

If the staff review as described in standard review plan Section 2.7 results in the acceptance of the disposal cell hydraulic conductivity, the following conclusions may be presented in the technical evaluation report:

The staff has completed its review of the disposal cell hydraulic conductivity at the

_____ uranium mill facility. This review included an evaluation using the review procedures in Section 2.7.2 and the acceptance criteria outlined in Section 2.7.3 of this standard review plan.

The licensee has acceptably evaluated the disposal cell cover materials hydraulic conductivity by providing a sufficient technical basis for the design K-value for the disposal cell. A field testing program adequate to verify the constructability of the disposal cell with a hydraulic design conductivity of $K<10^{-7}$ cm/sec is presented. The applicant followed an acceptable quality control program for the field testing to determine the hydraulic conductivity.

On the basis of the information presented in the application and the detailed review conducted of the disposal cell hydraulic conductivity at the _____ uranium mill facility, the NRC staff concludes that the disposal cell hydraulic conductivity and associated conceptual and numerical models provide an acceptable input to the demonstration of compliance with the following criteria in 10 CFR Part 40, Appendix A: Criterion 4(c), which provides requirements for the embankment and cover slopes for tailings and Criterion 6(1), which requires that impoundment designs provide reasonable assurance of control of radiological hazards to be effective for 1,000 years to the extent reasonably achievable, and in any case, for at least 200 years.

2.7.5   References

American Society for Testing and Materials Standards:

D 2434. "Test Method for Permeability of Granular Soils (Constant Head)."

D 3385. "Test Method for Infiltration Rate of Soils in Field Using Double-Ring Infiltrometers."

D 5093. "Test Method for Field Measurement of Infiltration Rate Using a Double-Ring Infiltrometer With a Sealed Inner Ring."

Benson, C.H. and D.E. Daniel. "Influence of Clods on Hydraulic Conductivity of Compacted Clay." *ASCE Journal of Geotechnical Engineering.* Vol. 116, No. 8. pp. 1,231–1,248. 1990.

NRC. NUREG/CR–3356, "Geotechnical Quality Control: Low-Level Radioactive Waste and Uranium Mill Tailings Disposal Facilities." Washington, DC: NRC. 1983.

———. Regulatory Guide 1.132, "Site Investigations for Foundations of Nuclear Power Plants." Rev. 1. Washington, DC: NRC, Office of Standards Development. March 1979.

Panno, S.V., et al. "Field-Scale Investigation of Infiltration Into a Compacted Soil Liner. *Ground Water.* Vol. 29, No. 6. pp. 914–921. 1991.

Rogowski, A.S. "Relationship of Laboratory- and Field-Determined Hydraulic Conductivity in Compacted Clay Layer." EPA/600/S2–90/025. Cincinnati, Ohio: Risk Reduction Engineering Laboratory. 1990.

3.0 SURFACE WATER HYDROLOGY AND EROSION PROTECTION

3.1 Hydrologic Description of Site

Criterion 1 of 10 CFR Part 40, Appendix A, addresses the general goals of siting and designing facilities to provide for permanent isolation of tailings, and minimizing the potential for dispersion by natural forces, without the need for active maintenance. Information presented in Section 3.1 will be used in later sections of this standard review plan to assess the ability of the site and the site design to meet this and other requirements of 10 CFR Part 40.

It is important to note that the siting criteria presented in 10 CFR Part 40, Appendix A are intended to apply to uranium mills that have not yet been constructed. For many, if not most, uranium mills, reclamation plans are developed for sites that have existed for several decades. In fact, many mills were producing uranium before the siting criteria were developed. Therefore, the staff concludes that Criterion 1 is more relevant to new facilities (or modifications to old facilities) than to facilities that existed before regulations were developed.

3.1.1 Areas of Review

The staff should review hydrologic site characterization information, including (1) identification of the relationships of the site to surface-water features in the site area and (2) identification of mechanisms, such as floods and dam failures, that may require special design features to be implemented. This review requires identification of the hydrologic characteristics of streams, lakes (e.g., location, size, shape, drainage area), and existing or proposed water control structures that may adversely affect the long-term stability of the site design features.

3.1.2 Review Procedures

The staff should evaluate the completeness of the information and data, by sequential comparison with information available from references. On the basis of the description of the hydrosphere (e.g., geographic location and regional hydrologic features), potential site flood mechanisms are identified. The information normally presented is not amenable to independent verification, except through cross-checks with available publications related to hydrologic characteristics of the site region and through observation during site visits.

The staff should also analyze geomorphic considerations, as described in Section 1 of this standard review plan. On the basis of these analyses, the staff should estimate the potential for geomorphic instability to occur and to have a significant effect on the ability of the site and its protective features to prevent flood intrusion and erosion over a long period of time. If geomorphic problems are identified, the staff should give particular attention to several areas of the design, depending on site conditions and potential for geomorphic changes to occur. These areas include the (1) apron and toe of the disposal cell, (2) intersection of natural gullies with erosion protection features, and (3) diversion channel outlets. A detailed discussion of the erosion protection design for these and other features is given in Section 3.4.2 of this standard review plan.

Surface Water Hydrology and Erosion Protection

3.1.3 Acceptance Criteria

The hydrologic description of the site will be considered acceptable if:

(1)   The description of structures, facilities, and erosion protection designs is sufficiently complete to allow independent evaluation of the impact of flooding and intense rainfall.

(2)   Site topographic maps are of good quality and of sufficient scale to allow independent analysis of pre- and post-construction drainage patterns.

(3)   The reclamation plan contains sufficient information for the staff to independently evaluate the hydraulic designs presented. In general, detailed information is needed for each method that is used to determine the hydraulic designs and erosion protection provided to meet NRC regulations. NUREG–1623 (NRC, 2002) discusses acceptable methods for designing erosion protection to provide reasonable assurance of effective long-term control and, thus, conform to NRC requirements. NUREG–1623 (NRC, 2002) also provides discussions and technical bases for use of specific criteria to meet the 1,000-year longevity requirement, without the use of active maintenance. Specific design methods are provided and form the primary basis for staff review of erosion protection designs.

3.1.4 Evaluation Findings

If the staff evaluation of hydrologic and hydraulic engineering aspects of the reclamation plan confirms that the information acceptably characterizes the site and the site design features, the following conclusions may be presented in the technical evaluation report:

The staff has completed its review of the flooding potential at the _____ uranium mill facility. This review included an evaluation using the review procedures in Section 3.1.2 and acceptance criteria outlined in Section 3.1.3 of this standard review plan.

On the basis of the information presented in the application and the detailed review conducted of the flooding potential for the _____ uranium mill facility, the NRC staff concludes that (1) the flood analyses and investigations adequately characterize the flood potential at the site, (2) the analyses of hydraulic designs are appropriately documented, and (3) the general reclamation plan with respect to surface-water hydrology and erosion considerations, represents a feasible plan, for complying with the requirements of 10 CFR Part 40, Appendix A. The characterization of flood potential and the documentation of the site design conform to the requirements of Criterion 1 of 10 CFR Part 40, Appendix A, which requires a design that provides for permanent isolation of tailings and minimizes disturbance and dispersion by natural forces.

3.1.5 References

NRC. NUREG–1623, "Design of Erosion Protection for Long-Term Stabilization." Washington, DC: NRC. 2002.

3.2 Flooding Determinations

3.2.1 Areas of Review

The staff should assess the flooding potential for the site, and should determine precipitation potential, precipitation losses, runoff response characteristics, and peak flow estimates for the probable maximum flood or project design flood (if a flood less than the probable maximum flood is used). The staff should review the following design analyses: (1) the analyses and justification for the use of a flood less than the probable maximum flood, if applicable; (2) the probable maximum precipitation potential and resulting runoff for site drainage and for drainage areas adjacent to the site; and (3) the modeling of physical rainfall and runoff processes to estimate flood conditions at the site.

The assessment of flooding also should include a review of possible geomorphic changes that could affect the erosion protection design for the site. As applicable, the staff should review the following: (1) identification of types of geomorphic instability; (2) changes to, and impacts associated with, flooding and flood velocities from geomorphic changes; and (3) mitigative measures to reduce or control geomorphic instability. This information must be reviewed to determine the acceptability of hydraulic engineering designs to mitigate the geomorphic conditions and to avoid the need for ongoing active maintenance.

The assessment of flooding should also include a review of potential dam failures, if upstream reservoirs exist. Peak water levels, flood routing procedures, and velocities should be reviewed in the determination of potential hazards because of failure of upstream water control structures from either seismic or hydrologic causes. If an existing analysis concludes that seismic or hydrologic events will not cause failures of upstream dams and produce the governing flood at the site, the analysis should be reviewed to verify that information that supports such a conclusion (e.g., record of contact with dam designers) is included. If an analysis is provided that concludes that a dam failure flood from a probable maximum flood or a seismically induced flood is the design-basis flood, the computations should be reviewed to verify that appropriate and/or conservative model input parameters have been used.

3.2.2 Review Procedures

The evaluation of flooding is, for review purposes, separated into two parts: (1) flooding on large adjacent streams, as applicable, and (2) localized flooding on drainage channels and protective features. The acceptability of using the probable maximum flood as the design flood event is presented in Section 2.2.1 of NUREG–1623 (NRC, 2002). The review procedure for evaluating a probable maximum precipitation/probable maximum flood event is outlined in Appendix D of NUREG–1623 (NRC, 2002). For large drainage areas, probable maximum flood estimates approved by the U.S. Army Corps of Engineers and found in published or unpublished reports of that agency, or generalized estimates, may be used instead of independent staff-developed estimates. The staff should also assess flood history in the site area by examining historic regional flood data. For many areas, historic flood peaks could be a small percentage of the probable maximum flood. If the historic maximum floods exceed or closely approximate the proposed probable maximum flood estimates, the staff should perform a detailed evaluation to determine the basis for the estimates. The staff should compare basin

Surface Water Hydrology and Erosion Protection

lag times, rainfall distributions, soil types, and infiltration loss rates to determine if there is a logical basis for the probable maximum flood values being less than historic floods. Without such estimates, the staff should generally use U.S. Army Corps of Engineers models to independently estimate probable maximum flood discharge and water levels at the site. If detailed computer models are used, the staff should review the adequacy of the various input parameters to the model, including, but not limited to, the following: drainage area, lag times and times of concentration, design rainfall, incremental rainfall amounts, temporal distribution of incremental rainfall, and runoff/infiltration relationships.

The staff should review the dam failure analyses presented in the reclamation plan or should independently estimate the peak flows at the site. Often, it may be much easier to perform simplified flood analyses assuming a dam failure, rather than detailed analyses of the seismic resistance of a dam. In such cases, the staff should review those simplified flood analyses using the procedures outlined in standard review plan Section 3.3.4.

The staff should evaluate the information presented in the reclamation plan using procedures found in Appendix C of NUREG–1623 (NRC, 2002) in those cases in which it is documented that it is impractical to design erosion protection features for an occurrence of the probable maximum flood. These procedures contain detailed information regarding justification of a stability period of less than 1,000 years. To assure that minimum NRC requirements are met, the staff should independently check and evaluate the ability of the design to resist such flood events.

In the detailed review of flooding, the staff should carefully consider the following factors that are important in determining a local probable maximum precipitation/probable maximum flood event:

- Determination of Design Rainfall Event. The staff should consult appropriate hydrometeorological reports and determine that correct values of the 1- and 6-hour probable maximum precipitation events, as applicable, have been given.

- Infiltration Losses. The staff should check calculations to verify that appropriate values of infiltration have been selected.

- Times of Concentration. The staff should verify that appropriate methods (depending on the slope, configuration, etc.) have been selected. The staff should independently verify that the methods selected compare reasonably well with various velocity-based methods.

- Rainfall Distributions. The staff should verify that the rainfall distributions (particularly the 2½-, 5-, and 15-minute distributions) compare well with the distributions suggested in Appendix D to NUREG–1623 (NRC, 2002).

For dam failures, the staff should review estimates of flood potential and water levels. Depending on the potential for flooding, the staff should verify that the dam failure analyses are either realistic or conservative by determining locations and sizes of upstream dams, assuming an instantaneous failure (complete removal) of the dam embankment, and computing the peak outflow rate.

If this simplified analysis indicates a potential flooding problem, the analysis may be repeated using more refined techniques, and the staff may request additional information and data. Detailed failure models, such as those of the Army Corps of Engineers and National Weather Service, will be used to identify the outflows, failure modes, and resultant water levels at the site.

Assessments of flooding will be used to determine the acceptability of hydraulic engineering design to avoid the need for ongoing active maintenance at the site.

If a flood less than a probable maximum flood can cause dam failure and is proposed as the design-basis flood, the staff should employ the review procedures outlined above to determine the impracticality of designing for a probable maximum flood and to determine the acceptability of the flood used.

3.2.3   Acceptance Criteria

The flooding determinations for the site will be considered acceptable if:

The designs conform to the suggested criteria in Appendix D to NUREG–1623 (NRC, 2002). NUREG–1623 (NRC, 2002) discusses acceptable methods for designing erosion protection to provide reasonable assurance of effective long-term control and to meet NRC requirements. It also presents discussions and technical bases for use of specific criteria to meet the 1,000-year longevity requirement without the use of active maintenance. Acceptable design methods are presented and form the primary basis for staff review of erosion protection designs. These methods were derived from regulatory requirements, other regulatory guidance, staff experience, and various technical studies.

Information pertinent to computation of the design flood is submitted in sufficient detail to enable the staff to perform an independent flood estimate, Specifically:

- Model input parameters are adequate.

- Staff and the reclamation plan estimates of flood levels and peak discharges are in agreement.

- Computational methods for design flood estimates are adequate.

"Worst conditions" postulated in the analysis of upstream dam failures are (1) an approximate 25-year flood on a normal operating reservoir pool level coincident with the dam-site equivalent of the earthquake for which the remedial action project is designed, (2) a flood of about one-half the severity of a probable maximum flood on a normal reservoir pool level coincident with the dam-site equivalent of one-half of the earthquake for which the remedial action project is designed; and (3) a probable maximum flood (or design flood) on a normal reservoir pool.

Surface Water Hydrology and Erosion Protection

Conditions 1 and 2 are applied when the dam is not designed with adequate seismic resistance; Condition 3 is applied when the dam is not designed to safely store or pass the design flood.

If the proposed design is based on less than a probable maximum flood event, the licensee offers reasonable assurance of conforming to the stability requirement of at least 200 years.

Dam failure analyses are either realistic or conservative, and include locations and sizes of upstream dams, instantaneous failure (complete removal) of the dam embankment, and compute the peak outflow rate.

3.2.4 Evaluation Findings

If the staff evaluation of hydrologic and hydraulic engineering aspects of the reclamation plan confirms that the assessments of flooding are acceptable, the following conclusions may be presented in the technical evaluation report.

The staff has completed its review of the flooding potential at the _____ uranium mill facility. This review included an evaluation using the review procedures in Section 3.2.2 and the acceptance criteria outlined in Section 3.2.3 of this standard review plan.

On the basis of information presented in the application and the detailed review conducted of the flooding potential for the _____ uranium mill facility, the NRC staff concludes that the flood analyses and investigations adequately characterize the flood potential at the site and that the surface water hydrology and flooding considerations represent a feasible plan for meeting the requirements of 10 CFR Part 40, Appendix A.

The mill tailings at the _____ uranium mill facility will be protected from flooding and erosion by an engineered rock riprap layer that has been designed in accordance with the guidance suggested by the staff. Flood analyses presented by the licensee demonstrate that this erosion protection is adequate, based on (1) selection of proper rainfall and flooding events; (2) selection of appropriate parameters for determining flood discharges; and (3) computation of flood discharges, using appropriate and/or conservative methods.

The licensee presented analyses to show that the site is located in an area rarely flooded by off-site floods and that it is protected from direct on-site precipitation and flooding. The erosion protection is large enough to resist flooding from the shallow depths and minimal forces of floods occurring from a probable maximum flood in the upstream drainage area. The staff therefore concludes that the erosion potential at the proposed site has been acceptably minimized, since any flooding at the site is mitigated by the erosion protection, and the forces associated with off-site floods are minimal. The staff also concludes that, because the rainfall and flooding events have very low probabilities of occurrence over a 1,000-year period, no damage to erosion protection is expected from these, or more frequent, events. Therefore, maintenance or repair of damage will not be necessary.

On the basis of the information presented in the application and the detailed review conducted of the flooding potential for the _____ uranium mill facility, the NRC staff concludes that the flood analyses contribute to meeting the following requirements of 10 CFR Part 40,

Appendix A: Criterion 1, requiring that erosion, disturbance, and dispersion by natural forces over the long term are minimized and that the tailings are disposed of in a manner that does not require active maintenance to preserve conditions of the site; Criterion 4(a), requiring that upstream rainfall catchment areas are minimized to decrease erosion potential and to resist floods that could erode or wash out sections of the tailings disposal area; Criterion 6(1), requiring that the design be effective for a period of 200–1,000 years; and Criterion 12, requiring that active maintenance is not necessary to preserve isolation.

3.2.5 References

NRC. NUREG–1623, "Design of Erosion Protection for Long-Term Stabilization." Washington, DC: NRC. 2002.

3.3 Water Surface Profiles, Channel Velocities, and Shear Stresses

3.3.1 Areas of Review

Depending on the type of computational models used, the staff should review the model, including the determination of flooding depths, channel velocities, and/or shear stresses used to determine riprap sizes needed for erosion protection. The staff should review the various detailed computations for each model and should review the acceptability of the input parameters to the model. The staff should estimate the flood levels, velocities, shear stresses, and magnitudes, as described below. The review should be oriented toward verifying that the site will not require ongoing active maintenance.

3.3.2 Review Procedures

Using the guidance presented in Appendix D to NUREG–1623 (NRC, 2002) the staff should verify that localized flood depths, velocities, and shear stresses used in models for rock size determination or soil cover slope analysis are acceptable. For off-site flooding effects, the staff should verify that computational models have been correctly and appropriately used and that the data from the model have been correctly interpreted. The staff should verify that acceptable models and input parameters have been used in all the various portions of the flood analyses and that the resulting flood forces have been adequately accommodated.

Staff estimates may be made independently from basic data, by detailed review and checking of the reclamation plan analyses, or by comparison with other estimates that have been previously reviewed in detail. The evaluation of the adequacy of the estimates is a matter of engineering judgment, and is based on the confidence in the estimate, the degree of conservatism in each parameter used in the estimate, and the relative sensitivity of each parameter as it affects the flood level, flood velocity, or design of the erosion protection.

The staff review should evaluate whether ongoing active maintenance will be required at the site.

3.3.3 Acceptance Criteria

The water surface profiles, channel velocities, and shear stresses calculated for the site will be considered acceptable if:

The proposed designs conform to the suggested criteria in Appendix D to NUREG–1623 (NRC, 2002). NUREG–1623 (NRC, 2002) discusses acceptable methods for designing erosion protection to provide reasonable assurance of effective long-term control and to comply with NRC requirements. This document also contains discussions and technical bases for use of specific criteria to meet the 1,000-year longevity requirement without the use of active maintenance. Specific design methods are presented, and reasonable similarity to these methods forms the primary basis for staff acceptance of erosion protection designs. Specifically:

- Localized flood depths, velocities, and shear stresses used in models for rock size determination or soil cover slope analysis conform to the guidance presented in Appendix D to NUREG–1623 (NRC, 2002).

- For off-site flooding effects, computational models have been correctly and appropriately used and the data from the models have been correctly interpreted.

- Acceptable models and input parameters have been used in all the various portions of the flood analyses and the resulting flood forces have been adequately accommodated.

3.3.4 Evaluation Findings

If the staff evaluation of hydrologic and hydraulic engineering aspects of the reclamation plan confirms that the assessments of flooding are acceptable, the following conclusions may be presented in the technical evaluation report:

The staff has completed its review of the flooding models at the _____ uranium mill facility. This review included an evaluation using the review procedures in Section 3.3.2 and the acceptance criteria outlined in Section 3.3.3 of this standard review plan.

On the basis of the information presented in the application and the detailed review conducted of the flooding models for the _____ uranium mill facility, the NRC staff concludes that flood velocities and forces associated with flooding at the site have been acceptably computed.

The mill tailings will be protected from flooding and erosion by an engineered rock riprap layer that has been designed in accordance with the guidance suggested by the staff. Flood analyses presented by the licensee demonstrate that adequate protection is provided by (1) selection of proper models to assess rainfall and flooding events, (2) selection of appropriate parameters for models for determining flood forces, and (3) computation of flood forces using appropriate and/or conservative methods.

The staff considers that the riprap layers proposed will not require active maintenance over the 1,000-year design life, because the licensee adopted models that conservatively compute flood forces used to design the erosion protection. Thus, the use of conservative design parameters will result in no damage to the erosion protection designed using those methods. The staff further concludes that the hydraulic design features are sufficient to protect the tailings from flood forces that are very large and have very low probabilities of occurrence over a 1,000-year period. Therefore, maintenance of the rock layers will not be necessary.

The staff concludes that the analyses and models used at the _____ uranium mill facility contribute to meeting the following requirements of 10 CFR Part 40, Appendix A: Criterion 1, requiring that erosion, disturbance, and dispersion by natural forces over the long term are minimized and that the tailings are disposed of in a manner that does not require active maintenance to preserve conditions of the site; Criterion 6(1), requiring the design to be effective for a period of 200 to 1,000 years; and Criterion 12, requiring that active ongoing maintenance is not necessary o preserve isolation of the tailings.

3.3.5 Reference

NRC. NUREG–1623, "Design of Erosion Protection for Long-Term Stabilization." Washington, DC: NRC. 2002.

3.4 Design of Erosion Protection

3.4.1 Areas of Review

Design details and analyses pertinent to the following aspects of erosion protection will be reviewed, as applicable:

(1) Erosion protection for slopes and channel banks to protect against flooding from nearby large streams.

(2) Erosion protection for the top and side slopes of the pile.

(3) Erosion protection for the apron/toe area of the side slope.

(4) Erosion protection for drainage and diversion channels, including channel outlets.

(5) Durability of the erosion protection.

Surface Water Hydrology and Erosion Protection

(6)   Construction considerations, including specifications, quality assurance programs, quality control programs, and inspection programs.

In Section 3.4.2.4 (below), sedimentation in diversion channels is also addressed. Criterion 4(f) of 10 CFR Part 40, Appendix A, suggests that deposition of sediment in impoundment areas should be considered for enhancing the cover thickness. The staff considers it important to differentiate between beneficial and detrimental sediment accumulations. For example, if sediment could be conveniently routed to the middle of an impoundment, without long-term erosion or ponding of runoff that could affect ground-water conditions, such deposition may enhance long-term cover thickness. However, this is difficult to actually achieve. The major problem with sediment is that it tends to accumulate in diversion channels that are constructed on relatively flat slopes. High-velocity runoff from steep slopes carries sediment into low-velocity diversion channels, and that sediment can eventually accumulate and completely block the channel. Thus, it can be seen that some sediment buildup is good and some is bad. The review should evaluate the need for ongoing active maintenance of the site.

3.4.2 Review Procedures

The staff should check the analyses in the reclamation plan or perform independent review analyses of floods, flood velocities, and rock durability according to the guidelines in Appendix D to NUREG–1623 (NRC, 2002). The following areas should be evaluated.

(1)   Banks of Natural Channels

       The staff should review designs for riprap to be placed on the side slopes of a reclaimed pile or on natural channel banks to protect against erosive velocities from floods on large rivers. Guidance is presented in Appendix D to NUREG–1623 (NRC, 2002) for assessing floods, determining input parameters to models, and determining riprap requirements.

(2)   Top Slope and Side Slopes

       The staff should review input parameters to calculations and models according to the recommendations given in Appendix D to NUREG–1623 (NRC, 2002) and referenced technical procedures. The staff should assess the design flow rate, the depth of flow, angle of repose, specific gravity, and other parameters. For both the top and side slopes, the rock sizes should be checked using the recently developed, simplified procedures discussed in NUREG–1623 (NRC, 2002).

(3)   Apron/Toe

       The design of the apron and toe is reviewed by verifying that several design features in this area have been properly designed, in accordance with the recommendations in NUREG–1623 (NRC, 2002).

       For the lower end of the side slope where it meets the toe, the staff should verify that proper consideration has been given to the potential occurrence of increased shear forces resulting from turbulence and energy dissipation produced by hydraulic jumps, when the flow transitions from supercritical to subcritical. The staff should verify that appropriate design criteria have been used to increase the rock size to account for the increased velocities or shear forces.

For the main area of the toe, the staff should assure that appropriate methods have been used to design the riprap, depending on the magnitude of the slope of the toe.

For the downstream end of the toe, the staff should verify that acceptable assumptions have been made regarding the assumed collapse of the rock into scoured areas to prevent gully intrusion. Flow concentrations, collapsed slopes, and computational models should be evaluated.

For the natural ground area at the downstream end of the toe, the staff should verify that appropriate methods have been used to compute scour depths and that natural erosion will not adversely affect long-term stability.

(4)   Diversion Channels

Using the criteria and guidance presented in Appendix D to NUREG–1623 (NRC, 2002), the staff should evaluate the design of diversion channels in several critical areas.

For the main channel area, the staff should verify that appropriate models and input parameters have been used to design the erosion protection. The staff should assure that flow rates, flow depths, and shear stresses have been correctly computed.

For the channel side slopes, the staff should verify that the side slopes are capable of resisting flow velocities and shear stresses from flows that occur directly down the side slope. This occurs often when diversion channels are constructed perpendicular to natural gullies (which discharge into the diversion channel). The shear forces in these locations often greatly exceed the forces produced by flows in the channel, particularly when the slope of the natural ground in the area is greater than the slope of the diversion channel.

For the outlet of the diversion channel, the staff should evaluate the design of erosion protection to assure that erosion in the discharge area (normally a natural gully, swale, or channel) has been adequately addressed. Designs similar to apron/toe designs should be evaluated to determine their resistance to erosion. Appendix D to NUREG–1623 (NRC, 2002) discusses acceptable methods for designing channel outlets.

For the entire length of the diversion channel, the staff should evaluate the effects of sediment accumulations on flow velocities, channel capacity, and need for increased rock size. Particular attention should be given to designs in which steep natural streams discharge into relatively flat diversion channels, greatly increasing the potential for blockage of the channel. Appendix E to NUREG–1623 (NRC, 2002) discusses acceptable methods for assessing sedimentation in diversion channels.

Surface Water Hydrology and Erosion Protection

(5)   Rock Durability

       The staff should review the results of durability testing of proposed rock sources to
       assure that durable rock will be used. Appendix D to NUREG–1623 (NRC, 2002)
       presents a detailed method for evaluating rock quality for various locations and
       applications. If durable rock is not available to the site, to the extent practical, the
       reviewer should review the alternative proposed by the applicant and the associated
       analysis to assure that the alternative provides reasonable assurance that the radon
       barrier will be effective for 1,000 years, to the extent reasonably achievable, and in any
       case, for at least 200 years.

(6)   Construction Considerations

       The staff should review the plans, specifications, inspection programs, and quality
       assurance/quality control programs to assure that adequate measures are being taken
       to construct the design features according to accepted engineering practices. The staff
       should compare the information presented with typical programs used in the
       construction industry. Appendix F to NUREG–1623 (NRC, 2002) contains examples of
       acceptable specifications and testing programs that were approved by the staff and
       actually applied at several sites.

(7)   The review shall specifically evaluate whether the erosion protection design is sufficient
       to avoid the need for ongoing active maintenance at the site.

3.4.3 Acceptance Criteria

The design of erosion protection for the site will be considered acceptable if:

The proposed designs conform to the suggested criteria in NUREG–1623 (NRC, 2002) .
NUREG–1623 (NRC, 2002) discusses acceptable methods for designing erosion protection to
provide reasonable assurance of effective long-term control and to comply with NRC
requirements. This document also contains discussions and technical bases for use of specific
criteria to meet the 1,000-year longevity requirement without the use of active maintenance.
Specific design methods are presented, and reasonable similarity to these methods forms the
primary basis for staff acceptance of erosion protection designs. NUREG–1623 (NRC, 2002)
updates and expands the final staff technical position (NRC, 1990).

If active maintenance is proposed as an alternative to the designs suggested above, such an
approach will be found acceptable if the following criteria are met:

(1)   The maintenance approach must achieve an equivalent level of stabilization and
       containment and protection of public health, safety, and the environment.

(2)   The licensee must demonstrate a site-specific need for the use of active maintenance
       and an economic benefit.

(3)   The licensee must provide funding for the maintenance by increasing the amount of the required surety. The staff should determine if the licensee's estimate of funding required for active maintenance is adequate. The licensee should also work with the long-term custodian to assess any additional funding requirements related to long-term surveillance and monitoring.

3.4.4   Evaluation Findings

If the staff evaluation of hydrologic and hydraulic engineering aspects of the reclamation plan confirms that the erosion protection designs are acceptable, the following conclusions may be presented in the technical evaluation report:

The staff has completed its review of the design of erosion protection at the _____ uranium mill facility. This review included an evaluation using the review procedures in Section 3.4.2 and the acceptance criteria outlined in Section 3.4.3 of this standard review plan.

On the basis of the information presented in the application and the detailed review conducted of the erosion protection features, the staff concludes that the designs are acceptable.

The mill tailings will be protected from flooding and erosion by an engineered rock riprap layer. The riprap has been designed in accordance with the guidance suggested by the NRC staff. The staff considers that erosion protection that meets that guidance will provide adequate protection against erosion and dispersion by natural forces over the long term. In addition to the adequacy of the flood analyses discussed in standard review plan Sections 3.2 and 3.3, the staff concludes that adequate erosion protection designs are provided by (1) use of appropriate methods for determining erosion protection needed to resist the forces produced by the design discharge, and (2) selection of a rock type for the riprap layer that will be durable and capable of providing the necessary erosion protection for a long period of time. Further, the staff considers that the riprap layers proposed will be durable over the 1,000-year design life, for the following reasons: (1) the rock proposed for the riprap layers was evaluated using rock quality procedures suggested by the staff and is not expected to deteriorate significantly over the 1,000-year design life; (2) the rock fragments are dense, resistant to abrasion, and free from cracks, seams, and other defects; and (3) during construction, the rock layers will be placed in accordance with appropriate engineering and testing practices, minimizing the potential for damage, dispersion, and segregation of the rock.

The riprap for the relatively flat top and side slopes is designed to be sufficiently large to minimize erosion potential. The rock will be capable of resisting flooding and erosion, depending on the slope selected. Thus, the staff concludes that the relatively steep slopes, with their corresponding rock designs, are acceptable.

On the basis of its review of the designs for the _____ uranium mill facility, the staff concludes that the hydraulic designs contribute to meeting the requirements of 10 CFR Part 40, Appendix A: (1) Criterion 1, requiring that erosion, disturbance, and dispersion by natural forces over the long term are minimized and that the tailings are disposed of in a manner that does not require active maintenance to preserve conditions of the site; (2) Criterion 4(c), requiring embankments and cover slopes to be relatively flat after stabilization to minimize erosion potential and to provide conservative factors of safety that ensure long-term stability; (3) Criterion 4(d), requiring that the rock cover reduces wind and water erosion to negligible levels, including consideration of such factors as the shape, size, composition, and gradation of the rock particles; (4) Criterion 4(f), requiring the design to promote deposition, where feasible; (5) Criterion 6(1), requiring the design to be effective for 200 to 1,000 years; and (5) Criterion 12, requiring that active on-going maintenance is not necessary to preserve isolation.

3.4.5 References

NRC. NUREG–1623, "Design of Erosion Protection for Long-Term Stabilization." Washington, DC: NRC. 2002.

————. "Design of Erosion Protection Covers for Stabilization of Uranium Mill Tailings Sites." Washington, DC: NRC. 1990.

3.5   Design of Erosion Protection Covers

3.5.1   Areas of Review

If a soil or vegetative cover is proposed, the following design details, calculations, and analyses will be reviewed:

(1)   Determination of allowable shear stresses and permissible velocities for the cover.

(2)   Determination of allowable shear stresses and permissible velocities for the cover in a degraded state, including the effects of fires, droughts, vegetation succession, and other impacts to the ability of the cover to function without maintenance.

(3)   Information on types of vegetation proposed and their abilities to survive natural phenomena.

(4)   Information, analyses, and calculations of input parameters to models used.

The review will consider whether the design of covers is sufficient to avoid the need for ongoing active maintenance at the site.

3.5.2   Review Procedures

If a soil cover is proposed, the staff should evaluate the design using the general criteria outlined in Appendix A to NUREG–1623 (NRC, 2002). Particular attention should be given to the input parameters to various models.

(1)   The staff should verify that the design flow rate includes an appropriate flow concentration factor that reflects consideration of settlement, soil removal by sheet flow and wind, degradation of the vegetation cover, intrusion of trees, blockage of flows by fallen trees, etc.

(2)   The staff should verify that estimates of Manning's "n" value correspond to the vegetation cover proposed and are proper for estimating allowable shear stresses and permissible velocities.

(3)   The staff should verify that appropriate values of allowable shear stresses and permissible velocities have been used and conservatively reflect potential changes that could occur to the cover over a long period of time as a result of fires, droughts, diseases, vegetation succession, or general cover degradation.

(4)   The staff should check analyses and/or independently calculate allowable slopes using several different methods and ranges of input parameters. Using a range of flow concentration factors, shear stresses, permissible velocities, "n" values, and models, the staff should check the sensitivity of the analyses and should verify that reasonable and appropriate values of input parameters have been selected.

If a sacrificial soil cover is proposed to meet the minimum 200-year stability requirement, the staff should check the calculations using Appendix B to NUREG–1623 (NRC, 2002) and the justification for reduction of the stability period using Appendix C to NUREG–1623 (NRC, 2002).

(5)   The reviewer shall determine whether the design is adequate to avoid the need for ongoing active maintenance at the site.

3.5.3 Acceptance Criteria

The design erosion protection covers for the site will be considered acceptable if:

The designs conform to the suggested criteria in NUREG–1623 (NRC, 2002). NUREG–1623 (NRC, 2002) discusses acceptable methods for designing erosion protection to provide reasonable assurance of effective long-term control and, thus, meet NRC requirements. This document also provides discussions and technical bases for use of specific criteria to meet the 1,000-year longevity requirement without the use of active maintenance. Specific acceptance criteria for many of the review areas are presented and form the primary basis for staff review of erosion protection designs. These criteria were derived from regulatory requirements, other regulatory guidance, staff experience, and various technical references.

If active maintenance is proposed as an alternative to the designs suggested above, such an approach will be found acceptable if the following criteria are met:

(1)   The maintenance approach must achieve an equivalent level of stabilization and containment and protection of public health, safety, and the environment.

(2)   The licensee must demonstrate a site-specific need for the use of active maintenance.

(3)   The licensee must provide funding for the maintenance by increasing the amount of the required surety. The licensee should also work with the long-term custodian to assess any additional funding requirements related to long-term surveillance and monitoring.

Surface Water Hydrology and Erosion Protection

3.5.4 Evaluation Findings

If the staff's evaluation of erosion protection covers confirms that the cover designs are acceptable, the following conclusions may be presented in the technical evaluation report:

The staff has completed its review of the design of erosion protection covers at the _____ uranium mill facility. This review included an evaluation using the review procedures in Section 3.5.2 and the acceptance criteria outlined in Section 3.5.3 of this standard review plan.

On the basis of its review, the staff concludes that the designs are acceptable and meet the requirements of 10 CFR Part 40, Appendix A.

The mill tailings will be protected from flooding and erosion by an engineered soil cover. The staff considers that a satisfactory cover will provide adequate protection against erosion and dispersion by natural forces over the long term. In addition to the adequacy of the flood analyses discussed in standard review plan Sections 3.2 and 3.3, the staff concludes that adequate cover designs are provided by:

(1)   Use of appropriate methods for determining cover slopes needed to resist the forces produced by the design discharge.

(2)   Selection of a cover that will be capable of providing the necessary erosion protection for a long period of time.

On the basis of the information presented in the application and the detailed review conducted of the erosion protection covers for the _____ uranium mill facility, the NRC staff concludes that the cover designs contribute to meeting the following requirements of 10 CFR Part 40, Appendix A: Criterion 1, requiring that erosion, disturbance, and dispersion by natural forces over the long term are minimized and that the tailings are disposed of in a manner that does not require active maintenance to preserve conditions of the site; Criterion 4(b), requiring siting and design such that topographic features provide good wind protection; Criterion 4(c), requiring that embankments and cover slopes are relatively flat after stabilization to minimize erosion potential and to provide conservative factors of safety; Criterion 6(1), requiring the design to be effective for 200 to 1,000 years; and Criterion 12, requiring that active ongoing maintenance is not necessary to preserve isolation.

3.5.5 References

NRC. NUREG–1623, "Design of Erosion Protection for Long-Term Stabilization." Washington, DC: NRC. 2002.

3.6   General References

American Nuclear Society. "American National Standard for Determining Design Basis Flooding at Power Reactor Sites." ANSI/ANS–2.8. 1981.

Bureau of Reclamation. "Comparison of Estimated Maximum Flood Peaks with Historic Floods." Washington, DC: U.S. Department of the Interior. 1986.

————. "Design of Small Dams." Second Edition. Washington, DC: U.S. Department of the Interior. 1973.

Chow, V.T. *Open Channel Hydraulics*. New York, New York: McGraw-Hill Book Company. 1959.

Crippen, J.R. and C.D. Bue. "Maximum Floodflows in the Conterminous United States." U.S. Geological Survey Water Supply Paper No. 1887. 1977.

Fread, D.L. "DAMBRK: The National Weather Service Dam-Break Flood Forecasting Model." Silver Spring, Maryland: National Weather Service. Continuously updated.

Henderson, F.M. *Open Channel Flow*. New York, New York: MacMillan Company. 1971.

Interagency Advisory Committee on Water Data, Hydrology Subcommittee Working Group on Probable Maximum Flood Risk Assessment. "Draft Report on the Feasibility of Assigning a Probability to the Probable Maximum Flood." June 1985.

Nelson, J.D., R.L. Volpe, R.E. Wardell, S.A. Schumm, and W.P. Staub. NUREG/CR–3397, ORNL–5979, "Design Considerations for Long-Term Stabilization of Uranium Mill Tailings Impoundments." Washington, DC: NRC. October 1983.

NRC. "Design of Erosion Protection Covers for Stabilization of Uranium Mill Tailings Sites." Final Staff Technical Position. Washington, DC: NRC. 1990.

————. "Standard Format and Content for Documentation of Remedial Action Selection at Title I Uranium Mill Tailings Sites." Washington, DC: NRC. February 1989.

Simons, D.B. and F. Senturk. *Sediment Transport Technology*. Fort Collins, Colorado: Water Resources Publications. 1977.

Stephenson, D. "Rockfill Hydraulic Engineering Developments in Geotechnical Engineering." No.27. Elsevier Scientific Publishing Company. 1979.

Temple, D.M., et al. "Stability Design of Grass-Lined Open Channels." Agricultural Handbook Number 667. U.S. Department of Agriculture. 1987.

U.S. Army Corps of Engineers. "Flood Hydrograph Package." HEC–1. Davis, California: Hydrologic Engineering Center. Continuously updated.

————. "Water Surface Profiles." HEC–2. Davis, California: Hydrologic Engineering Center. Continuously updated.

Surface Water Hydrology and Erosion Protection

———. "Reservoir System Operation for Flood Control." HEC–5. Davis, California: Hydrologic Engineering Center. Continuously updated.

———. "Hydraulic Design Criteria." Continuously updated and revised.

———. "Runoff from Snowmelt." EM1110–2–1406. January 1980.

———. "Additional Guidance for Riprap Channel Protection." ETL1110–2–120. May 1971.

———. "Hydraulic Design of Flood Control Channels." EM1110–1601. July 1970.

———. "Policies and Procedures Pertaining to Determination of Spillway Capacities and Freeboard Allowances for Dams." EC1110–2–27. February 1968.

———. "Hydraulic Design of Spillways." EM1110–2–1603. March 1965.

———. "Standard Project Flood Determinations." EM1110–2–1411. March 1965.

———. "Interior Drainage of Leveed Urban Areas: Hydrology." EM1110–2–1410. May 1965.

———. "Routing of Floods through River Channels." EM1110–2–1408. March 1960.

———. "Backwater Curves in River Channels." EM1110–2–1409. December 1959.

———. "Flood Hydrograph Analysis and Computations." EM1110–2–1405. August 1959.

———. "Stone Protection." CE 1308. January 1948.

Weather Bureau. "Hydrometeorological Reports." (Now U.S. Weather Service). Hydrometeorological Branch: Nos. 43, 49, and 55.

4.0 PROTECTING WATER RESOURCES

The protection of water resources is a process that encompasses two distinct strategies. The first strategy is to prevent the spread of contaminants from disposal and processing sites into ground water or surface water. This strategy requires the staff to ensure that operations and decommissioning are conducted in such a manner as to minimize threats to ground water.

The second strategy is to mitigate the threat to public health and the environment from contaminants that have already been mobilized—particularly through ground water pathways—before initiation of decommissioning activities. This strategy applies only to those sites where ground-water contamination already exists and requires staff to review existing or proposed ground-water restoration activities to ensure that they will result in compliance with regulatory requirements. The NRC exercises exclusive, pre-emptive jurisdiction over all radiological and non-radiological ground-water contaminants from uranium mill tailings facilities, in accordance with Commission direction in Staff Requirements Memorandum SECY 099-277 (NRC, 2000).

Use of this chapter should be tailored to the specific situation and phase of operation at each site. The reviewer will select and emphasize the various aspects of the areas covered by this standard review plan chapter. The judgment on the areas to be given attention during the review is to be based on the specific licensee submittal being reviewed, an inspection of the material presented, and prior knowledge of the site and its operating history.

This chapter presents a standard approach for reviewing, evaluating, and documenting the evaluation findings for issues pertaining to water resource protection during the various phases of the license termination process at licensed uranium mill sites. Review of information concerning the protection of water resources shall be coordinated with the evaluation of the site stratigraphy, structural and tectonic information, and surface water and erosion protection information as described in standard review plan Chapters 1.0 and 3.0, respectively. Review procedures in this chapter pertain to the following four types of documents that are submitted for review by the staff:

(1)   Licensees submit reclamation plans to obtain approval of surface reclamation and decontamination work, including stabilization of mill tailings, and elimination (or isolation) of present or potential contaminant sources.

(2)   Licensees submit corrective action plans during operations or during the license termination process to obtain approval of ground-water restoration strategies at sites where ground-water contamination has been detected.

(3)   Licensees submit ground-water restoration completion reports to confirm that the ground-water quality will remain stable after ground-water restoration strategies have been implemented and that ground-water protection standards have been correctly established.

(4)   Long-term custodians submit long-term surveillance plans to describe the ground-water monitoring activities that will be implemented by the custodian.

Protecting Water Resources

The ultimate objective of the review is to determine if the proposed reclamation plans and corrective action plans will result in long-term compliance with 10 CFR Part 40, Appendix A. As stated in 10 CFR Part 40, Criterion 5, "Criteria 5A-5D and new Criterion 13 incorporate the basic ground-water protection standards imposed by the Environmental Protection Agency in 40 CFR Part 192, Subparts D and E (48 FR 45926; October 7, 1983), which apply during operations and prior to the end of closure. Ground-water monitoring to comply with these standards is required by Criterion 7A." To meet this regulatory objective, the following issues must be evaluated:

(1)   Site characterization.

(2)   Ground-water protection standards.

(3)   Hazard and as low as is reasonably achievable assessment for alternate concentration limits, as defined by 10 CFR Part 40, Appendix A, Criteria 5B(5) and 5B(6).

(4)   Ground-water corrective action and monitoring plans.

Accordingly, this chapter contains a section for each of these areas. Discussions in this chapter incorporate acceptable practices from all previous staff technical positions and guidance documents pertaining to uranium mill tailings reclamation. This standard review plan supercedes those documents. The NRC exercises exclusive jurisdiction over all radiological and non-radiological ground-water contaminants from uranium mill tailings facilities.

4.1   Site Characterization

4.1.1   Areas of Review

The staff should review the characterization information, given the circumstances and life cycle of a particular site, and the nature of the document under review (reclamation plan, corrective-action plan). The staff should also evaluate regional and site-specific hydrologic information related to both the former processing site and the proposed disposal site if they are different. The hydrologic information should include both surface-water and ground-water systems, along with any interrelations among those systems. Complete site characterization should include or reference the following:

(1)   Site background data that include descriptions of:

     (a)   The site history of mining and/or milling operations.

     (b)   Surrounding land and water uses.

     (c)   Site meteorological data.

(2) Ground-water and surface-water hydrology data, including:

   (a) Descriptions of hydrogeology and ground-water conditions.

   (b) Estimation of hydraulic and transport properties for each hydrogeologic unit.

   (c) Descriptions of surface-water hydrology and estimations of ground-water and surface-water interactions.

   (d) Assessment of potential for flooding and erosion.

(3) Information concerning geochemical conditions and water quality, including:

   (a) Identification of constituents of concern.

   (b) Determination of background ground-water quality.

   (c) Confirmation of proper statistical analysis.

   (d) Delineation of the nature and extent of contamination.

   (e) Identification of contaminant source terms.

   (f) Characterization of subsurface geochemical properties.

   (g) Identification of attenuation mechanisms and estimation of attenuation rates.

(4) Human health and environmental risk evaluations, including:

   (a) Radiological risks.

   (b) Non-radiological risks.

   (c) A summary of risk evaluations from the site environmental report.

4.1.2 Review Procedures

The level of effort necessary to adequately characterize a particular site depends on site-specific circumstances. For example, if a particular site has no ground-water contamination and tailings are disposed off site, there will be very little need for detailed site characterization in support of water resources protection. Conversely, at a site with an existing source of ground-water contamination, the site characterization must be sufficient to support selection of cleanup strategies and to determine the level of risk to human health and the environment.

Protecting Water Resources

There is not a single acceptable approach to conducting a site characterization, because the appropriate level of site characterization is specific to the methods of tailings disposal and ground-water corrective action selected for a particular site. As such, the reviewer should:

(1)   Thoroughly evaluate the characterization information using the acceptance criteria in standard review plan Section 4.1.3, but reserve final judgment until all sections of the application have been reviewed.

(2)   Assess whether the level of detail and technical merit of the characterization are sufficient to support the proposals, assumptions, and assertions in the application that are used to demonstrate regulatory compliance.

4.1.3 Acceptance Criteria

Knowledge of the site is needed to evaluate the existing and potential contamination. This characterization information shall include a description of activities and physical properties that may affect water resources at the mill site. The site characterization will be acceptable if it meets the following criteria:

(1)   It contains a description of the site that is sufficient to assess the environmental impact the former mill site may have on the surrounding area; the populations that may be affected by such impacts; and meteorological conditions that may act to transport contaminants off site. An acceptable site description will contain the following specific information:

      (a)   A site history that includes:

            (i)   A list of the known leaching solutions and other chemicals used in the milling process and their relative quantities in mill wastes. The list should also identify any constituent listed in 10 CFR Part 40, Appendix A, Criterion 13, that may have been disposed of in the tailings pile.

            (ii)   A description of the wastes generated at the site during milling operations, waste discharge locations, types of retaining structures used (e.g., tailings piles, ponds, landfills), quantities of waste generated, and a chronology of waste management practices.

            (iii)   A summary of the known impacts of the site activities on the hydrologic system and background water quality.

            (iv)   If applicable, descriptions of any human activities or natural processes unrelated to the milling operation that may have altered the hydrogeologic system. Such human activities include ground-water use, crop irrigation, mine dewatering, ore storage, municipal waste land filling, oil and gas development, or exploratory drilling. Natural processes include geothermal springs, natural concentration of soluble salts by evaporation, erosion processes, and ground-water/surface-water interactions.

(b) Information pertaining to surrounding land and water uses that includes:

    (i) A general overview of water uses, locations, quantities of water available, and the potential uses to which the quality of water is suited>

    (ii) Definitions of the class-of-use category for each water source (e.g., drinking water, agricultural, livestock, limited use).

    (iii) Identification of potential receptors of present or future ground-water or surface-water contamination.

    (iv) Descriptions of non-mill-related human activities or natural processes that may affect water quality or water uses (e.g., oil and gas development, municipal waste landfills, crop irrigation, drought, and erosion).

    Human water consumption is not the only water use that must be considered in the review. Any use that may bring someone into contact with the contaminated water must be considered when evaluating health hazards. For example, non-potable, radon-contaminated water piped to a public lavatory could pose a substantial health hazard.

(c) Sufficient meteorologic data for the region, including rainfall, temperature, humidity and evaporation data in sufficient detail to assess projected water infiltration through the disposal cell.

Monthly averages are an acceptable means of presenting general meteorological conditions; however, the reviewer shall ensure that extreme weather conditions are adequately described.

(2) The ground-water and surface-water hydrology is described adequately to support modeling predictions of likely contaminant migration paths; selection of monitor well locations; and, when ground-water contamination exists, selection of a restoration strategy. The following specific information is provided to support these objectives:

(a) A description of hydrogeologic units that may affect transport of contaminants away from the site via ground-water pathways, including:

    (i) Hydrostratigraphic cross sections and maps are included to delineate the geometry, lateral extent, thickness, and rock or sediment type of all potentially affected aquifers and confining zones beneath the processing and disposal sites of such quality and quantity to support a technically defensible interpretation.

    (ii) The hydrogeologic units that constitute the uppermost aquifer (where regulatory compliance will be evaluated) are identified. The uppermost aquifer is the geologic formation nearest the natural ground surface that is an aquifer, as well as lower aquifers that are hydraulically interconnected with this aquifer within the facility property boundary.

(iii)   If local perched aquifers are found at the site, their presence is noted. These formations may cause contaminated water to be diverted around monitoring systems, or may be improperly interpreted as the uppermost aquifer. Any saturated zone created by uranium or thorium recovery operations would not be considered an aquifer unless the zone is or potentially is (1) hydraulically interconnected to a natural aquifer, (2) capable of discharge to surface water, or (3) reasonably accessible because of migration beyond the vertical projection of the boundary of the land transferred for long-term government ownership and care in accordance with 10 CFR Part 40, Appendix A, Criterion 11.

(iv)   Unsaturated zones, through which contaminants may be conveyed to the water-bearing units, are described. This information is adequate to support the assumptions used in estimating the source term for contaminant transport pathways. This information includes identification of potential preferential flow pathways that are either natural (e.g., buried stream channels), or man-made (e.g., abandoned wells or mine shafts).

(v)   Information is presented on geologic characteristics that may affect ground-water flow beneath the former mill site. Examples of pertinent geologic characteristics include identification of significant faulting in the area, fracture and joint orientation and spacing for the underlying bedrock, and geomorphology of soil and sedimentary deposits (e.g., fluvial, glacial, or volcanic deposits).

(vi)   Hydraulic-head contour maps, of both local and regional scale, for the uppermost aquifer and any units connected hydraulically beneath the site are sufficient to determine hydraulic gradients, ground-water flow direction, and proximity to offsite ground-water users. These maps are based on static water level observations at onsite and regional wells. Several measurements are taken at each observation well (American Society for Testing and Materials Standards D 4750, D 5092, D 5521, D 5787, and D 5978). These measurements are sufficiently spaced in time to capture water-level fluctuations caused by seasonal changes or local pumping of ground water. Enough observation wells are sampled to produce an adequate water elevation contour map. The appropriate number of wells is dependent on the size of the site and the choice of contour interval. However, as a rough estimate, there is at least one observation well for each contour line on the map. A more detailed contour map (small contour interval) is produced for the site and surrounding properties. The level of detail used for the regional contour map may be limited by the number of observation wells available offsite. The reviewer shall bear in mind that calculations of hydraulic gradients from hydraulic head contour maps are only rigorously valid for horizontal flow in aquifers.

(b)   Estimations of hydraulic and transport properties of the underlying aquifer

Hydrogeologic parameters used to support the choice of a ground-water restoration strategy or to demonstrate compliance include hydraulic conductivity, saturated thickness of hydro geologic units, hydraulic gradient, effective porosity, storage coefficient, and dispersivity. The reviewer shall consider the influence of each of these parameters on evaluating compliance with standards established pursuant to 10 CFR Part 40, Appendix A, and determine whether estimates for each parameter are reasonably conservative, based on the data provided.

(i)   Hydraulic conductivity and storage coefficients are determined by conducting aquifer pump tests on several wells at the site. Pump test methods that are consistent with American Society for Testing and Materials standards for the measurement of geotechnical properties and for aquifer hydraulic tests are considered acceptable by the NRC. These American Society for Testing and Materials Standards are D 4044, D 4050, D 4104, D 4105, D 4106, D 4630, D 5269, D 5270, D 5472, D 5473, D 5737, D 5785, D 5786, D 5850, D 5855, D 5881, and D 5912. Any other peer-reviewed method or commonly accepted practice for aquifer parameter estimation may be used. When curve fitting is used to analyze pump test data, deviations of observation data from ideal curves are explained in terms of likely causes (e.g., impermeable or recharge boundaries, leaky aquitards, or heterogeneities). When average hydraulic parameters are reported, the reviewer shall consider that many hydrogeologic parameters, including hydraulic conductivity, typically exhibit a log-normal distribution. Consequently, the geometric mean may be more representative of the overall conditions within a unit than the arithmetic mean.

(ii)   Horizontal components of hydraulic gradient are estimated by measurement of the distance between contour intervals on hydraulic head contour maps. Vertical components of hydraulic gradient are estimated from head measurements in different aquifers or at different depths in the same aquifer.

(iii)   Generally, analyses considering steady-state conditions are acceptable unless site conditions indicate otherwise. If transient conditions are modeled, storage coefficients estimated from standard tests indicated in (i) above are used.

(iv)   If contaminant transport is modeled, then longitudinal and transverse dispersivity values are either obtained from a tracer test or conservative values based on published literature are used. Because dispersivities depend on the size of the modeled region, the reviewer shall carefully

4-7

compare the values for dispersivity used in the licensee transport modeling with those values cited in survey studies such as Gelhar, et al. (1992), and verify that they represent conservative estimates for the site.

(c)   Estimation of ground-water/surface-water interactions at sites with nearby streams, rivers, or lakes.

The locations of surface-water bodies that are connected to the site ground-water flow system are identified. Surface-water elevations shall be used to help describe the site ground-water flow system if a stream or other surface-water body discharges into or drains the site ground-water flow system. Another acceptable approach is to evaluate hydraulic head contour based on data from monitor wells in the vicinity of streams.

(3)   Geochemical conditions and water quality are characterized sufficiently to:

(a)   Identify the constituents of concern.

Any chemical or radiological constituent that is reasonably expected to be in or derived from the tailings is a potential constituent of concern. 10 CFR Part 40, Appendix A, Criterion 13 provides a non-inclusive list of constituents of concern which standards must be set and complied with. Criterion 13 also provides flexibility to add constituents on a case-by-case basis.

Table 4.1.3-1 presents a list of constituents commonly associated with uranium mill tailings.[1] This list is based on a chemical survey performed by NRC staff at 17 licensed mill tailings sites.

Most of the constituents in 10 CFR Part 40, Appendix A, Criterion 13 are organic compounds that are not normally associated with uranium milling processes. The expected presence of organic compounds is assessed from knowledge of the chemicals used during the milling process or other materials that may have been disposed of in the tailings. If there is no record of organic compounds used in the process, screening tests for volatile and semi-volatile organic compounds are performed to confirm the absence of organic compounds in the tailings and groundwater.

---

[1]Smith, R.D. "Memorandum (February 9) Sampling of Uranium Mill Tailings Impoundments for Hazardous Constituents." 1987.

Table 4.1.3-1. Common Uranium Mill Chemical Constituents	
Inorganic Constituents	**Organic Constituents**
Arsenic	Carbon Disulfide
Barium	Chloroform
Beryllium	Diethyl Phthalate
Cadmium	2—Butanone
Chromium	1,2—Dichloroethane
Cyanide	Naphthalene
Lead	
Mercury	
Molybdenum	
Net Gross Alpha*	
Nickel	
Radium-226 and -228	
Selenium	
Silver	
Thorium-230	
Uranium	
* Excluding Radon, Radium, and Uranium	

Staff may require the addition of constituents associated with the milling process that are not specifically listed in 10 CFR Part 40, Appendix A, Criterion 13, to ground-water monitoring programs. These constituents may be added on a case-by-case basis, if they are capable of posing a substantial present or potential hazard to human health or the environment. If the staff requires a constituent to be added to the list in Criterion 13, the NRC must establish an associated compliance limit for each added constituent at a level that will be protective of human health and the environment.

Some constituents which typically do not present a hazard to human heath and the environment may pose such a hazard to some specific human or environmental populations, under site-specific circumstances. As an example, three constituents associated with uranium mill tailings may be candidates for site-specific evaluations during licensing reviews and potential NRC regulation on a case-by-case basis, under specific circumstances. Illustrative constituents, circumstances, and potential harm are tabulated in Table 4.1.3-2.

The above examples are not all inclusive. The reviewer should examine these and other constituents that produce similar potential harm under specific circumstances. Non-radiological constituents that degrade the water quality and produce and impact on the designated water use beyond the proposed long-term

Table 4.1.3-2. Non-Radiological Ground-Water Constituents That May Produce Harm		
Constituent	**Exposure Circumstance**	**Potential Harm**
Sodium	Drinking water pathway, human exposure	Some segments of the human population with elevated blood pressure may be sensitive to sodium intake above a recommended limit. The EPA added sodium to its Drinking Water Contaminant Candidate List for further evaluation.
Sulfate	Drinking water pathway, human exposure	Some segments of the human population are sensitive to elevated sulfate in drinking water, which can produce osmotic diarrhea. The EPA added sulfate to its Drinking Water Contaminant Candidate List for further evaluation.
Ammonia, Ammonium ion	Surface water pathway, aquatic organism exposure	Various aquatic species are sensitive to ammonia levels as low as 0.38 mg/L. These levels are far lower than exposure limits that would produce an adverse impact to human populations.

The above examples are not all-inclusive. The reviewer should examine these and other constituents that produce similar potential harm under specific circumstances. Non-radiological constituents that degrade the water quality and produce and impact on the designated water use beyond the proposed long-term care boundary must also be evaluated to determine whether they should be included in the license. The reviewer should consult with the appropriate non-Agreement State agency on the designated water use for the ground-water resource an any numerical limits the State has determined to be a hazard.

Close coordination with the State may be needed to determine the need for including such constituents in the license, along with evaluating the benefits and costs of potential mitigative measures.

In identifying additional constituents, the staff should ensure that any additions are made based on a sound technical and regulatory basis. Examples of sound technical bases are the following:

(1) NRC and the U.S. Environmental Protection Agency agree to use one federal contact with a licensee, which is NRC. This approach requires NRC to include some constituents in its licenses, that are not normally licensed by the NRC.

(2) Trends in ground-water contamination show that after several years of decreases in the level of contamination, the level of contamination is beginning to rise again.

(3)Surrogate parameters that cover a family of constituents show an increase in concentration in ground water. Therefore, the staff may require licensees to monitor for all constituents found in that family.

(4) Some constituents used in the milling process, but not listed in Criterion 13, may pose a hazard to some specific human and environmentally sensitive populations, under site-specific circumstances, including degradation of a designated water use beyond the proposed long-term care boundary.

Even if the criteria for identifying a constituent of concern are met, NRC may still decide to exclude certain constituents on a site-specific basis if it can be shown that the constituents are not capable of posing a substantial present or potential hazard to human health or the environment. In considering such exclusions, the reviewer must consider potential adverse effects on ground-water quality and hydraulically connected surface-water quality. NRC may decide to exclude a constituent if the dissolved concentration of the constituent in the tailing fluids is equal to or less than the concentration of that constituent in the background water quality. Alternately, NRC may decide to exclude a constituent if the dissolved concentration of the constituent in the tailing fluids is equal to or less than the maximum value for ground-water protection listed in 10 CFR Part 40, Appendix A, Table 5C.

4-11

New constituents of concern should be added in a timely manner. This is done either at the time the corrective action plan is accepted for review, or at some time during the lifetime of the corrective action plan. New constituents will not be required at the time of the license termination monitoring submittal, unless the one-time, pre-termination ground-water sampling identifies constituents at concentrations that pose a hazard to human health and the environment. The reviewer should consult Appendix E (Section E3.3.2(1)) for those sites nearing license termination, regarding the one-time, pre-termination ground-water sampling and analysis.

(b)   Present a determination of background (baseline) water quality.

Background water quality is defined as the chemical quality of water that would be expected at a site if contamination had not occurred from the uranium milling operation.

Water quality data available from studies conducted in conjunction with initial licensing for operation of the facility are used to establish the background. If constituents of concern identified by NRC were not sampled in the original background monitoring program, the licensee should have conducted additional sampling to establish background levels. When adequate site-specific baseline data cannot be obtained for identified constituents of concern, samples of adjacent, and up-gradient, uncontaminated water are taken as proxies to onsite baseline samples.

To determine acceptability of the determination of background water quality, the following information is provided:

(i)   Maps are of sufficient detail and legibility to show the background monitoring locations.

(ii)   Descriptions of sampling methods, monitoring devices, and quality assurance practices are provided. Examples of acceptable methods are those that are consistent with American Society for Testing and Materials Standards D 4448, D 4696, and D 4840. Other methods, if used, are properly referenced and justified.

(iii)   When they exist, zones of differing background water quality are delineated. The possible causes of these differing water quality zones are discussed (e.g., changes from geochemically oxidizing to reducing zones in the aquifer; changes in rock type across a fault boundary).

(iv)   A table for each zone of distinct water quality, listing summary statistics (i.e., mean, standard deviation, and number of samples) for baseline water quality sampling for each constituent of concern, is provided.

4-12

(v)   A pre-operational monitoring program has been in place for 1 year, consistent with the requirements of 10 CFR Part 40, Appendix A, Criterion 7. Samples are taken at least monthly under this program. However, it is unlikely that mills in existence prior to the ground-water compliance provisions of 10 CFR Part 40, Appendix A, will have one full year of monthly baseline data from a pre-operational monitoring program.

Alternatively, background water quality may already be defined by a condition in the license. If this is the case, background limits for a ground-water protection standard have already been identified, and the reviewer should rely on those along with any constituents and standards listed in Criterion 5(c) as the regulatory limits applicable to this site.

(c)   Confirm the proper use of statistical techniques for assessing water quality.

Statistical hypothesis testing methods used for (i) establishing background water quality; (ii) establishing ground-water protection standards for compliance monitoring; (iii) determining the extent of ground-water contamination; and (iv) establishing the ground-water cleanup goals, are described in American Society for Testing and Materials Standard D 6312.

(d)   Define the extent of contamination.

A hazardous constituent is defined in 10 CFR Part 40, Appendix A, Criterion 5B(2), as a constituent that meets all three of the following tests:

(i)   The constituent is reasonably expected to be in or derived from the byproduct material in the disposal area.

(ii)   The constituent has been detected in the ground water in the uppermost aquifer.

(iii) The constituent is listed in Part 40, Appendix A, Criterion 13.[2]

For each hazardous constituent, the licensee determines the extent of contamination in ground water at the site. Ground-water contamination at uranium mill sites is usually limited to the uppermost aquifer. Maps showing the locations of sampling wells should be included, along with a discussion of sampling practices. The most useful way to present this information is on a map showing concentration contours for each hazardous constituent and water surface elevation contours. In this manner, the reviewer readily examines the size, shape, source, and direction of movement.

---

[2]Including a constituent which may pose a hazard to some specific human or environmentally sensitive populations, under site-specific circumstances, and can, therefore, be added to the list in Criterion 13 on a case-by-case basis.

The extent of contamination is delineated in three dimensions. This typically involves drilling a number of characterization wells and determining whether the water quality in each of these wells meets background water quality or whether the ground water is contaminated. It may not be necessary to sample all hazardous constituents to delineate the extent of contamination. Two or three indicator parameters (e.g., total dissolved solids, and chloride) might be selected. These indicators should be conservative—meaning that they are neither reactive, nor are they easily sorbed to soil—so that they provide a good indication of the maximum extent of contamination.

The transition from contaminated to uncontaminated ground water is often gradual. Thus, difficulty arises in determining where the contaminated water ends and the background water begins. The background sample data provide the easiest means for comparison of characterization well measurements to background measurements for the indicator parameters. Statistical methods described in American Society for Testing and Materials Standard D 6312 are suitable for determining whether contaminant concentrations exceed background levels.

Complications in delineating the extent of contamination arise at sites that have zones of differing water quality, or where onsite background water quality is not properly determined before discovery of ground-water contamination. Where zones of differing water quality are present, the reviewer shall verify that characterization wells are compared with the background sample from the appropriate water quality zone. Where onsite background water quality has not been properly determined, then upgradient or offsite samples are obtained.

The reviewer shall verify that the licensee has presented the following information to support determining the extent of contamination.

(i)   A map or maps showing the distribution of surface wastes and contaminated materials at and near the site.

(ii)   A map or maps showing the approximate shape and extent of ground-water contamination (e.g., concentration contour maps for indicator parameters in ground water).

(iii)   Identification of any offsite sources of water contamination or other factors that may have a bearing on observed water quality.

(e)   Properly estimate the source term.

Existing sources of ground-water contamination are defined in terms of location and rate of entry into the subsurface. At some sites, the contaminant sources have been effectively eliminated through stabilization or removal of tailings piles. However, residual sources may still exist in contaminated subsurface soils at the site. For ground-water contamination that originates from an onsite tailings pile, the source term is determined based on the chemical properties of the leachate and the rate at which leachate is released from the disposal area. The level of review given to source term calculations is commensurate with the overall importance of source term estimations to the selection of the restoration strategy.

    (i)   Source terms are reasonably correlated to the history of ore processing. All facilities from which leakage can occur are identified. Leaking constituents are identified based on the nature of the processing fluids. The volume of leakage is estimated in a realistic yet conservative manner. This can be done using water balance calculations, infiltration modeling, or seepage monitoring approaches.

    (ii)   When geochemical models are used to predict the fate and transport of existing contamination where the original source has been eliminated, the distribution of each hazardous constituent in place is taken as the source term.

(f)   Characterize the subsurface geochemical properties.

To effectively model the fate and transport of contaminants in ground water, it is important to characterize the geochemical properties of the natural waters and the aquifer mineralogy. Characterization of the underlying lithologies includes measurements of buffering capacity, total organic carbon, cation exchange capacity, and identification of the clay mineralogy. The general chemical characteristics of fluids within the lithologies are described by measurements of pH, temperature, dissolved oxygen, redox potential, buffering capacity, and the concentrations of major ions and trace metals.

    (i)   Aquifer geochemistry data are adequate to model the attenuation of contaminants. The values of the geochemical parameters used in transport models are justified. Acceptable parameter estimation methods are direct measurement, use of a conservative bounding estimate, reference to literature values for similar aquifer conditions, and laboratory studies of aquifer materials.

(g)   Identify contaminant attenuation mechanisms.

The major attenuation mechanisms that work to mitigate the effects of ground-water contamination are dilution in surrounding ground water, sorption of contaminants to the soil matrix, and immobilization of contaminants from geochemical and biochemical reactions.

    (i)   Claims that contamination is reduced by dilution are supported by a sufficient technical basis. There are two mechanisms for dilution of a contaminant plume in ground water: dispersion and mixing. Dispersion is a process whereby contaminant plumes tend to spread out and become less concentrated as they move away from the source. Mixing is the result of uncontaminated water being added to the ground-water system through natural recharge, injection, or upward movement of water from underlying aquifers, which reduces the concentration of contaminants. Estimation of surface recharge or upward flow through leaky aquitards is either established from field measurements, or through use of conservative assumptions are used.

(ii)   The values of sorption coefficients are based on the nature of the constituent and site-specific geochemical conditions. The degree of sorption of contaminants to the soil matrix depends on the affinity of each constituent for the soil in a particular aquifer. Constituents that carry a positive charge, as do most trace metals in solution, are good candidates for cation exchange adsorption to clay and oxide surfaces. However, because surface charges of clays and oxides decrease with decreasing pH, the reviewer shall carefully examine claims of attenuation from cation exchange under low pH conditions. Organic contaminants tend to be hydrophobic and are strongly attenuated in soils that have high organic carbon content. Most contaminant fate and transport models quantify the affinity of contaminants for soil by use of a distribution coefficient or K_D. Batch or column equilibria experiments, using representative leachate and soil samples, are performed to support estimations of K_D for each hazardous constituent.

(iii)   Attenuation from geochemical or biochemical equilibrium reactions is estimated by use of acceptable modeling software packages such as MINTEQA2 (Allison, et al., 1991) and PHREEQE (Parkhurst, et al., 1980). However, these packages are limited in that they do not consider transport of contaminants. Thus, results are only valid for reactions within a confined space (e.g., within the disposal cell). More recently developed reactive transport models [e.g., PHREEQC Version 2 (Parkhurst and Appello, 1999)] are also acceptable for constructing a geochemical model for the site. The reviewer shall determine that all model input parameters have sufficient technical bases and represent reasonably conservative estimations. Additionally, conclusions drawn from such models are supported by field observation; that is, they are consistent with site characterization data.

(iv)   At sites from which the contamination source has been effectively eliminated, monitoring data are used to assess attenuation of contaminants. If the contaminant source has been eliminated by surface reclamation, changes in the nature and extent of contamination over time are monitored. In such situations the center of mass of the contaminant plume moves along the direction of ground-water flow. The effects of dispersion are also observable over time as a decrease in peak concentrations near the center of the contaminant plume and a lateral spreading of the plume. If significant precipitation or adsorption is occurring, it is reflected in a decrease in the mass of contaminants in the aqueous phase.

4.1.4 Evaluation Findings

If the staff review, as described in standard review plan Section 4.1, results in the acceptance of the site characterization, the following conclusions may be presented in the technical evaluation report:

The staff has completed its review of the site characterization at the _____ uranium mill facility. This review included an evaluation using the review procedures in Section 4.1.2 and the acceptance criteria outlined in Section 4.1.3 of this standard review plan.

The licensee has presented an acceptable history of the site, including (1) a description of leaching solutions and other chemicals used in the process and their relative quantities; (2) a description of (a) the wastes generated at the site during the milling process, and (b) the waste handling facilities; (3) a summary of the known impact of site activities on the hydrologic system and water quality; and (4) a description of activities unrelated to uranium milling that may have altered the hydrologic system.

The licensee has presented acceptable information pertaining to the surrounding land and water use including (1) an overview of water uses, quantity available, and potential uses to which the water is suited; (2) definitions of the class-of-use category of each water source; (3) identification of potential receptors of ground-water or surface-water contamination; (4) assessment of variations in dilution effects of stream flow on contaminants; and (5) assessments of the effects of meteorological conditions on erosion, infiltration, and water-table elevation.

The licensee has presented acceptable meteorologic data, including (1) wind speed and direction, (2) rainfall, (3) evaporation data (4) temperature, and (5) humidity, to allow an evaluation of potential impacts of the meteorologic conditions on disposal cell performance.

The ground-water and surface-water hydrology is adequately described, including (1) geometry, lateral extent, and thickness of potentially affected aquifers and confining units; (2) a determination of which aquifers constitute the uppermost aquifer where regulatory compliance will be evaluated; (3) descriptions of the unsaturated units that convey hazardous constituents to the water-bearing units; (4) maps of acceptable detail showing the relative dimensions and locations of hydrogeologic units that have been impacted by milling activities; (5) information on geologic characteristics that may affect ground-water flow beneath the site; and (6) hydraulic head contour maps of both local and regional scale for the uppermost aquifer beneath the site.

The estimation of hydraulic and transport properties is acceptable and includes (1) hydraulic conductivity and storage coefficients determined by conducting aquifer pump tests on several wells; (2) determination of hydraulic gradients using hydraulic head contour maps; (3) calculations of storage coefficients, as applicable; and (4) longitudinal and transverse dispersivities, as appropriate. The evaluation of ground-water/surface-water interactions with nearby streams, rivers, or lakes is acceptable.

Geochemical conditions and water quality are adequately analyzed, including identification of constituents of concern that are reasonably expected to be derived from the tailings. Each constituent of concern is found in 10 CFR Part 40, Appendix A, Table 5C or 10 CFR Part 40,

Protecting Water Resources

Appendix A, Criterion 13, or has been included as a specific condition in the license. The licensee has made an acceptable determination of baseline water quality, including (1) maps of appropriate scale and legibility; (2) descriptions of sampling methods, monitoring devices, and quality assurance practices; (3) where applicable, delineation of zones of differing water quality and their possible origin; and (4) a table of summary statistics for each zone of differing quality. The applicant has presented an acceptable delineation of the extent of contamination supported by appropriate samples, maps of surface wastes and contaminated materials, maps of the approximate shape and extent of ground-water contamination, and identification of any off-site sources of water contamination. The description of the source term is acceptable and includes not only mill tailings constituents but those contaminants that might mobilize by contact with tailings leachate.

The characterization of the subsurface geochemical properties is acceptable. Attenuation mechanisms have been described, including the technical bases for determining that contamination will be reduced by dilution, sorption on the soil matrix, or geochemical or biochemical reactions. The licensee has presented direct measurements in support of attenuation of contaminants where the source has been eliminated by surface reclamation.

On the basis of the information presented in the application and the detailed review conducted of the site characterization for the _____ uranium mill facility, the NRC staff concludes that the information is acceptable and is in compliance with the following criteria in 10 CFR Part 40, Appendix A: Criterion 5B, which requires the NRC to establish a list of hazardous constituents, concentration limits, a point of compliance, and a compliance period; Criterion 5C, which provides a table of concentration limits for certain constituents when they are present in ground water above background concentrations; Criterion 5E, which requires licensees conducting ground-water protection programs to consider the use of bottom liners, recycle of solutions and conservation of water, dewatering of tailings, and neutralization to immobilize hazardous constituents; Criterion 5F, which requires that, where ground-water impacts caused by seepage are occurring at an existing site, action be taken to alleviate the conditions that lead to seepage and restore ground-water quality. Technical specifications for the seepage control system must be established, and a quality assurance, testing and inspection program must be established to insure the specifications are met. Criterion 5G, which requires that licensees/operators perform site characterization in support of a tailings disposal system proposal; Criterion 5H, which requires steps be taken during stockpiling of ore to minimize penetration of radionuclides into underlying soils; Criterion 7 which requires a year of monitoring prior to mill operations; Criterion 7A, which requires three types of monitoring systems: detection, compliance, and corrective action; and Criterion 13, which provides a list of hazardous constituents that must be considered when establishing the list of hazardous constituents in ground water at any site.

4.1.5 References

Allison, J.D., D.S. Brown, and K.J. Novo-Gradac. "MINTEQA2/PRODEFA2, A Geochemical Assessment Model for Environmental Systems: Version 3.0 User's Manual." EPA Publication EPA/600/3–91/021. Washington, DC: EPA. 1991.

American Society for Testing and Materials Standards:

D 4044, "Standard Test Method for (Field Procedure) for Instantaneous Change in Head (Slug) Tests for Determining Hydraulic Properties of Aquifers."

D 4050, "Standard Test Method (Field Procedure) for Withdrawal and Injection Well Tests for Determining Hydraulic Properties of Aquifer Systems."

D 4104, "Standard Test Method (Analytical Procedure) for Determining Transmissivity of Nonleaky Confined Aquifers by Overdamped Well Response to Instantaneous Change in Head (Slug Tests)."

D 4105, "Standard Test Method (Analytical Procedure) for Determining Transmissivity and Storage Coefficient of Nonleaky Confined Aquifers by the Modified Theis Nonequilibrium Method."

D 4106, "Standard Test Method (Analytical Procedure) for Determining Transmissivity and Storage Coefficient of Nonleaky Confined Aquifers by the Theis Nonequilibrium Method."

D 4448, "Standard Guide for Sampling Ground-water Monitoring Wells."

D 4630, "Standard Test Method for Determining Transmissivity and Storage Coefficient of Low-Permeability Rocks by In Situ Measurements Using the Constant Head Injection Test."

D 4750, "Standard Test Method for Determining Subsurface Liquid Levels in a Borehole or Monitoring Well (Observation Well)."

D 4840, "Standard Guide for Sampling Chain-of-Custody Procedures."

D 5092, "Standard Practice for Design and Installation of Ground Water Monitoring Wells in Aquifers."

D 5269, "Standard Test Method for Determining Transmissivity of Nonleaky Confined Aquifers by the Theis Recovery Method."

D 5270-96, "Standard Test Method for Determining Transmissivity and Storage Coefficient of Bounded, Nonleaky, Confined Aquifers."

D 5472, "Standard Test Method for Determining Specific Capacity and Estimating Transmissivity at the Control Well."

Protecting Water Resources

D 5473, "Standard Test Method for (Analytical Procedure for) Analyzing the Effects of Partial Penetration of Control Well and Determining the Horizontal and Vertical Hydraulic Conductivity in a Nonleaky Confined Aquifer."

D 5521, "Standard Guide for Development of Ground-Water Monitoring Wells in Granular Aquifers."

D 5737, "Standard Guide for Methods for Measuring Well Discharge."

D 5785, "Standard Test Method for (Analytical Procedure for) Determining Transmissivity of Confined Nonleaky Aquifers by Underdamped Well Response to Instantaneous Change in Head (Slug Test)."

D 5786, "Standard Practice for (Field Procedure) for Constant Drawdown Tests in Flowing Wells for Determining Hydraulic Properties of Aquifer System."

D 5787, "Standard Practice for Monitoring Well Protection."

D 5850, "Standard Test Method for (Analytical Procedure for) Determining Transmissivity, Storage Coefficient, and Anisotropy Ratio from a Network of Partially Penetrating Wells."

D 5855, "Standard Test Method for (Analytical Procedure for) Determining Transmissivity and Storage Coefficient of a Confined Nonleaky or Leaky Aquifer by Constant Drawdown Method in a Flowing Well."

D 5881, "Standard Test Method for (Analytical Procedure) Determining Transmissivity of Confined Nonleaky Aquifers by Critically Damped Well Response to Instantaneous Change in Head (Slug)."

D 5912, "Standard Test Method for (Analytical Procedure for) Determining Hydraulic Conductivity of an Unconfined Aquifer by Overdamped Well Response to Instantaneous Change in Head (Slug)."

D 5978, "Standard Guide for Maintenance and Rehabilitation of Ground-Water Monitoring Wells."

D 6312, "Standard Guide for Developing Appropriate Statistical Approaches for Ground-Water Detection Monitoring Programs."

Gelhar, L.W., C. Welty, and K.R. Rehfeldt. "A Critical Review of Data on Field-scale Dispersion in Aquifers." *Water Resources Research*. Vol. 28, No. 7. 1992.

NRC. "Concurrent Jurisdiction of Non-Radiological Hazards of Uranium Mill Tailings." Staff Requirements Memorandum to SECY–99–0277. Washington DC: NRC; Office of the Executive Director of Operations. August 11, 2000.

———. NUREG–1748, "Environmental Review Guidance for Licensing Actions Associated with NMSS Programs." Washington, DC: NRC, Office of Nuclear Material Safety and Safeguards. 2001

Parkhurst, D.L. and A.A.J. Appello. "User's Guide to PHREEQC (Version 2)—A Computer Program for Speciation, Batch-Reaction, One-Dimensional Transport, and Inverse Geochemical Modeling." 99-4259. Washington, DC: U.S. Geological Survey. 1999.

Parkhurst, D.L., Thorstensen, and L.N. Plummer. "PHREEQEM-A Computer Program for Geochemical Calculations." U.S. Geological Survey Water Resources Investigation 80-96. 1980.

4.2   Ground-Water Protection Standards

4.2.1   Areas of Review

Ground-water protection standards are established for each hazardous constituent. A hazardous constituent is defined in 10 CFR Part 40, Appendix A, Criterion 5B(2), as a constituent that meets all three of the following tests:

(1)   The constituent is reasonably expected to be in or derived from the byproduct material in the disposal area.

(2)   The constituent has been detected in the ground water in the uppermost aquifer.

(3)   The constituent is listed in 10 CFR Part 40, Appendix A, Criterion 13.

Even when constituents meet the three aforementioned tests, the Commission may exclude a detected constituent from the set of hazardous constituents, on a site-specific basis, if it finds that the constituent is not capable of posing a substantial present or potential hazard to human health or the environment. In deciding whether to exclude constituents, the considerations identified in 10 CFR Part 40, Appendix A, Criterion 5B(3), must be considered. In addition, as required by 10 CFR Part 40, Appendix A, Criterion 5B(4), any underground sources of drinking water and aquifers exempted by the EPA will be considered. Relevant EPA guidance is presented in 40 CFR 144.7, 144.3, and 146.4. The staff should review the technical basis the licensee has presented for the following elements of acceptable ground-water protection standards:

(1)   The list of hazardous constituents.

(2)   A description of the point of compliance.

(3)   Ground-water protection standards for hazardous constituents that may be either:

    (a)   Commission-approved background concentrations.

    (b)   Maximum concentration limits.

Protecting Water Resources

(c)   Alternate concentration limits.

The staff should also review additional ground-water protection standards that contain provisions for ground-water protection dealing with the design of surface impoundments and tailings disposal cells. Evaluation of disposal system performance is addressed in standard review plan Section 4.3.3.

4.2.2   Review Procedures

The reviewer should examine the ground-water protection standards to verify that they have been defined consistent with the acceptance criteria in standard review plan Section 4.3.2. Specifically, the reviewer should reference the existing license or:

(1)   Verify that the licensee has identified all constituents of concern that are present in the tailings leachate.

(2)   Verify that the point of compliance has been properly delineated.

(3)   Evaluate whether the proposed concentration limits for each ground-water protection standard are within a range that is reasonably expected to represent background concentrations; or, if any alternate concentration limits are proposed, verify that the appropriate evaluations have been presented in accordance with Criterion 5(B)(6) of 10 CFR Part 40, Appendix A.

4.2.3   Acceptance Criteria

Ground-water protection standards establish a concentration limit for each hazardous constituent at the point of compliance. The development of ground-water protection standards will be acceptable if it meets the following criteria:

(1)   Hazardous constituents are identified using the definition given in 10 CFR Part 40, Appendix A, Criterion 5(B).

(2)   A point of compliance is established in accordance with 10 CFR Part 40, Appendix A, Criterion 5B(1).

The point of compliance is the location at which the ground water is monitored to determine compliance with the ground-water protection standards. The objective in selecting the point of compliance is to provide the earliest practicable warning that the impoundment is releasing hazardous constituents to the ground water. The point of compliance must be selected to provide prompt indication of ground-water contamination on the hydraulically downgradient edge of the disposal area. The point of compliance is defined as the intersection of a vertical plane with the uppermost aquifer at the hydraulically downgradient limit of the waste management area.

The "uppermost aquifer" is defined in 10 CFR Part 40, Appendix A, as "the geologic formation nearest the natural ground surface that is an aquifer, as well as lower aquifers that are hydraulically interconnected with this aquifer within the facility's property boundary." Therefore, a proper selection of the point of compliance includes identification of point of compliance locations in the aquifer nearest to the ground surface, as well as other aquifers that are hydraulically interconnected with that aquifer, as warranted by site-specific conditions.

When tailings are disposed of on site, the NRC generally interprets the downgradient limit of the waste management area to be the edge of the reclaimed tailings side slopes. However, it is not recommended that licensees be required to compromise the cover integrity to install monitoring wells at the actual edge of the reclaimed tailings.

(3)   A concentration limit is specified for each of the hazardous constituents. Those limits may be:

(a)   Commission-Approved Background Concentrations.

10 CFR Part 40, Appendix A, requires that the Commission-approved background concentration be the concentration limit, except for constituents listed in Table 5C of 10 CFR Part 40, Appendix A, which, if present in excess of background, are subject to the respective maximum concentration limits listed in Table 5C.

Proper statistical methods, such as those discussed in American Society for Testing and Materials Standard D 6312, are used to determine the expected range of naturally occurring background (baseline) concentrations for each constituent of concern. Acceptable statistical techniques are also presented in Haan (1977) and Hirsch, et al. (1992).

(b)   Maximum Concentration Limits

The respective values given in the table in paragraph 5C of 10 CFR Part 40, Appendix A, must not exceed if the constituent is listed in the table and if the background level of the constituent is below the value listed. Note that the U.S. EPA has revised some of these limits under the Safe Drinking Water Act, therefore, for risk assessments used for an alternate concentration limit proposal, where a drinking water exposure pathway is estimated, the reviewer should refer to the most recent Safe Drinking Water Act maximum concentration limits.

(c)   Alternate Concentration Limits

Alternate concentration limits are established on a site-specific basis, provided it can be demonstrated that (i) the constituents will not pose a substantial present or potential hazard to human health or the environment, as long as the alternate concentration limits are not exceeded and (ii) the alternate concentration limits are as low as is reasonably achievable, considering practicable corrective actions. Licensees are required to implement detection monitoring programs to detect and identify site-specific hazardous constituents, and compliance monitoring programs to verify compliance with the established site-specific

Protecting Water Resources

standards for individual constituents. Standard review plan Sections 4.3.3 and 4.4.3 contain acceptance criteria for determining potential hazards, and for "as low as is reasonably achievable" demonstrations, respectively.

When an applicant proposes alternate concentration limits, the reviewer should recognize that additional site characterization may be necessary to demonstrate the potential risk to human health and the environment is acceptable. Typically, long-term ground-water monitoring will be required to assure that human health and the environment are protected.

4.2.4   Evaluation Findings

If the staff review, as described in standard review plan Section 4.2, results in the acceptance of the site ground-water protection standards, the following conclusions may be presented in the technical evaluation report:

The staff has completed its review of the ground-water protection standards at the _____ uranium mill facility. This review included an evaluation using the review procedures in Section 4.2.2 and the acceptance criteria outlined in Section 4.2.3 of this standard review plan.

The licensee has acceptably identified the hazardous constituents and has established acceptable concentration limits and cleanup standards. Established background levels are acceptable. Acceptable statistical methods have been used to establish the concentration limits. If alternate concentration limits have been requested, the licensee has acceptably supported the request with appropriate data and calculations. The licensee has established an acceptable point of compliance at the edge of the tailings impoundment on the downgradient direction of hydraulic flow.

On the basis of the information presented in the application and the detailed review conducted of the ground-water protection standards for the _____ uranium milling facility, the NRC staff concludes that the information is acceptable and is in compliance with the following criteria in 10 CFR Part 40, Appendix A: Criterion 5B, which requires the NRC to establish a list of hazardous constituents, concentration limits, a point of compliance, and a compliance period and, which allows use of alternate concentration limits under certain conditions; Criterion 5C, which provides a table of secondary concentration limits for certain constituents when they are present in ground water above background concentrations; Criterion 7A, which requires three types of monitoring systems: detection, compliance, and corrective action; and Criterion 13, which provides a list of hazardous constituents that must be considered when establishing the list of hazardous constituents in ground water at any site.

4.2.5   References

American Society for Testing and Materials Standards

D 6312, "Standard Guide for Developing Appropriate Statistical Approaches for Ground-Water Detection Monitoring Programs."

Haan, C.T. *Statistical Methods in Hydrology*. Iowa State University Press. 1977.

Hirsch, R.M., D.R. Helsel, T.A. Cohn, and E.J. Gilroy. *Statistical Analysis of Hydrologic Data, Handbook of Hydrology*. D.R. Maidment, ed. New York, New York: McGraw-Hill, Inc. 1992.

4.3 Hazard Assessment, Exposure Assessment, Corrective Action Assessment, and Compliance Monitoring for Alternate Concentration Limits

4.3.1 Areas of Review

The staff shall review the following elements of an alternate concentration limit application to determine regulatory compliance with 10 CFR Part 40, Appendix A, Criterion 5B(6):

(1)   Hazardous constituent(s) and the associated human and environmental risks of those constituent(s), including human cancer risk and environmental hazards. Characterization of the hazardous constituent source term and the extent of ground-water contamination.

(2)   Assessment of hazardous constituent transport in the ground water and hydraulically connected surface waters, and its adverse effects on water quality, including present and potential health and environmental consequences of exposure to the identified hazards.

(3)   A demonstration that a hazardous constituent concentration will not pose substantial present or potential hazard to human health or the environment at the point of exposure, and that the proposed alternate concentration limit is as low as is reasonably achievable, considering practicable corrective actions.

In addition, the implementation of the proposed alternate concentration limit and any modifications to the compliance monitoring program must be reviewed.

4.3.2 Review Procedures

Appendix K provides a description of a standardized content and format for an alternate concentration limit application. Strict conformance with the standard format is not a requirement, but review effectiveness and efficiency should be enhanced through its use. The proposed alternate concentration limit should be supported by a hazard assessment, an exposure assessment, and a corrective action assessment. Although separately listed, the information contained within each assessment should be integrated with the information that is developed in the subsequent assessment, so that all three assessments will collectively support the proposed alternate concentration limit. Appropriate portions of standard review plan Section 4.1 should also be consulted when performing a review of an alternate concentration limit application. The reviewer shall examine the provided information and assessments to determine that:

Protecting Water Resources

(1)   The source term has been adequately characterized to provide a realistic estimate of the types, characteristics, and the release rates of constituents of concern, which have been, or are expected to be, released to the ground water.

(2)   The risks and hazards that the released or potentially mobile constituents of concern may have on human health and the environment have been identified.

(3)   The extent of existing and potential ground-water contamination from the source term has been defined. The rates and directions of hazardous constituent migration and transport in the ground water and hydraulically connected surface waters have been determined. The point of compliance and point of exposure are identified.

(4)   The pathways for human and environmental exposure to the hazardous constituent(s) have been identified, and exposure magnitudes and consequences, including the human cancer risk, have been acceptably evaluated.

(5)   The proposed alternate concentration limit(s) for the point of compliance will result in a hazardous constituent concentration that is protective of human health and the environment at the point of exposure. The attenuation capacity of the aquifer between the point of compliance and the point of exposure has been adequately considered. There will be no adverse effects on the ground-water or on surface-water quality that would cause unacceptable health or environmental hazards at or beyond the point of exposure.

The applicant's assessment of ground-water corrective action alternatives shall be reviewed in conjunction with the hazard and exposure assessments. Previous, current, and potential future practicable corrective actions shall be evaluated against the costs and benefits of those actions to determine if the proposed alternate concentration limit is as low as is reasonably achievable. This demonstration should identify alternate corrective actions; assess their technical feasibility for implementation, and evaluate all associated costs and benefits of those corrective actions.

An alternate concentration limit must be protective of human health and the environment at the point of exposure, which is any location at or beyond the long-term care site boundary. A proposed alternate concentration limit that is not protective of human health and the environment, by itself, will not comply with the regulatory requirements for an alternate concentration limit. In this instance the applicant must submit the proposed numerical limit and any additional measures to protect human health and the environment to the Commission as an alternative to the specific requirements of 10 CFR Part 40, Appendix A, Criterion 5B (6), as permitted by section 84(c) of the Atomic Energy Act of 1954, as amended. The NRC staff will evaluate these alternatives on a case-by-case basis and determine the acceptability of the proposed alternative. A proposed alternate concentration limit that is not protective of human health and the environment and does not include additional alternate measures to provide such protection is not acceptable.

If the proposed alternate concentration limit is found acceptable, the compliance monitoring program must be evaluated before the license is terminated to determine that it is properly designed and implemented to ensure ground-water constituent concentrations in excess of the approved alternate concentration limit will be detected and that human health and the environment will be protected. Standard review plan Section 4.4 should be consulted.

4.3.3   Acceptance Criteria

The hazard assessment, exposure assessment, and corrective action assessment supporting a proposed alternate concentration limit will be acceptable if they meet the following criteria:

4.3.3.1 Hazard Assessment

The hazard assessment identifies all potential constituents of concern at a site. A potential constituent of concern is any compound that may be in or could be derived from the uranium mill tailings at a licensed site. A non-inclusive list of constituents of concern is in 10 CFR Part 40, Appendix A, Criterion 13. The risks and hazards to human health and the environment associated with those constituents are also identified and evaluated to determine whether an alternate concentration limit should be proposed for those constituents, if the subsequent exposure assessment concludes that an exposure is reasonably likely. Once a constituent of concern is released into the ground-water, it is classified as a hazardous constituent for the purpose of regulatory compliance, as described in 10 CFR Part 40, Appendix A, Criterion 5B(2). The hazard assessment should include the following:

(1)   The source term for all constituents of concern is adequately characterized and the extent of existing and potential future ground-water contamination is determined.

The source term characterization provides relevant information about the facility including: (a) the mechanical and chemical processes used to recover the uranium, (b) the types and quantities of the reagents used in milling, (c) the physical and chemical composition of the uranium-bearing ore, and (d) the historical and current waste and tailings management practices. This information is considered, in conjunction with the physical and chemical composition of the tailings and the type and distribution of existing contaminants, such as the location of waste discharge points, retaining structures for wastes, and waste constituents. The source characterization should provide reliable estimates of the release rates of hazardous constituents as well as constituent distributions.

(2)   The assessment identifies and evaluates the risks and hazards presented by the identified constituents of concern, including the human cancer risk caused by exposure to radioactive and non-radioactive constituents of concern, along with other health hazards that may be caused by the chemical toxicity of those constituents. The human cancer risk should be evaluated for individual constituents, including radioactive and carcinogenic chemicals, and compared with the maximum permitted risk level. The health effects of non-radioactive and non-carcinogenic constituents that are chemically toxic will be evaluated considering their risk-specific dose levels. It may be necessary to calculate a hazard index using the reference doses for those chemicals that have threshold effects. The hazard index is the ratio of calculated intake to the reference dose. An acceptable hazard index must be less than one. These evaluations distinguish between the health effects associated with threshold and non-threshold

constituents. Mutagenic, teratogenic, and synergistic effects are considered in the analysis, if applicable, based on toxicological testing, or structure-activity relationships.

The following additional information on constituent properties is provided, as applicable: (a) density, solubility, valence state, vapor pressure, viscosity, and partitioning coefficient; (b) presence and effects of complexing ligands and chelating agents that may enhance constituent mobility; (c) potential for a constituent to degrade because of biological, chemical, and physical processes; and (d) constituent attenuation properties, considering such processes as ion exchange, sorption, precipitation, dissolution, and ultrafiltration. This information would also be applied in the exposure assessment.

(3)   The assessment provides a reasonably conservative or best estimate of the potential health effects caused by human exposure to the hazardous constituent. The potential health effects for each constituent with a proposed alternate concentration limit must be identified, and related to appropriate exposure limits and dose-response relationships from available literature or databases. Sources of exposure limit and dose-response information include the EPA's maximum concentration limits for drinking water, reference doses, or risk-specific doses. Reference doses are the amounts of chemically toxic constituents to which humans may be daily exposed without suffering adverse effects.

Risk-specific doses are the amounts of proven or suspected carcinogenic constituents to which humans can be daily exposed, without increasing their risk of contracting cancer above a specified risk level. The reference dose and risk-specific dose assessment assume a human mass of 70 kg [154 lb] and consumption of 2 liters of water per day [0.53 gal/day]. More stringent criteria may apply if sensitive populations are exposed to hazardous constituents. Maximum concentration limits, reference doses, and/or risk-specific doses, can be used to show compliance with the risk level and hazard indexes. The technical basis for a risk assessment can be based on the dose-response relationships described in the scientific literature searches or toxicological research, in the absence of applicable maximum concentration limits, reference doses, or risk-specific doses. The exposure analysis should distinguish between threshold (toxic) and non-threshold (carcinogenic) effects associated with human exposure, as well as teratogenic, fetotoxic, mutagenic, and synergistic effects.

The maximum concentration limits, reference doses, and risk-specific doses for most hazardous constituents can be obtained from the EPA (http://www.epa.gov), the Agency for Toxic Substances and Disease Registry (http://www.atsdr.cdc.gov/atsdrhome.html), or other government institutions and universities. Effects from radioactivity can be obtained from the International Commission on Radiological Protection, and the National Council on Radiation Protection and Measurement.

Previously established and documented health-based constituent concentration limits are used in the hazard assessment as a basis for proposing alternate concentration limit values at specific sites.

(4)   The assessment identifies and evaluates the risks posed by the hazardous constituents to environmental populations. Adverse effects on aquatic and terrestrial wildlife, plants,

agricultural crops, livestock, and physical structures should be considered. Examples of these adverse effects are: (a) contaminant-induced changes in the biota, (b) loss or reduction of unique or critical habitats, and (c) jeopardy to endangered or threatened species. The NRC must initiate special consultation with the U.S. Fish and Wildlife Service, in accordance with 50 CFR Part 17, if endangered or threatened species occur on the site or could be impacted by site activities. NUREG–1748 (NRC, 2001) should be consulted for initiating this consultation.

Similar to the human risk evaluation, the environmental risk evaluation identifies any acute and sub-chronic effects on environmental populations caused by exposure to the hazardous constituents. Bioaccumulation and food chain interactions are considered when evaluating adverse effects. A comparison of the estimated constituent concentrations to the appropriate federal or State water-quality criteria should be part of the evaluation of potential effects on aquatic wildlife.

When appropriate, the hazard assessment considers potential damage to physical structures such as foundations, underground pipes, and roads. The applicant should demonstrate that the forecasted constituent concentrations will not result in any significant degradation or loss of function, as a result of contamination exposure. As an example, excessive concentrations of dissolved salts could result in accelerated corrosion of underground utility piping.

4.3.3.2 Exposure Assessment

The purpose of the exposure assessment is to evaluate the potential harm to human health and the environment from the hazards identified in the hazard assessment. The exposure assessment takes into account site-specific circumstances that may reduce or enhance the potential for exposure to hazardous constituents. This assessment identifies and evaluates hazardous constituent exposure pathways, and provides forecasts of human and environmental population responses, based on the projected constituent concentrations, dose levels, and available information on the radiological and chemical toxicity effects of the constituents. The assessment also addresses the underlying assumptions, variability, and uncertainty of the projected health and environmental effects. Exposure pathways should be identified and evaluated using water classification and water use standards, along with existing and anticipated water uses. Agricultural, industrial, domestic, municipal, environmental, and recreational water uses should also be considered, as they pertain to the site and surrounding areas. The exposure assessment must provide adequate information regarding potential effects on ground-water resources, and the above water uses, to support NRC's environmental review under 10 CFR Part 51. NUREG–1748 (NRC, 2001) should be consulted for the details of this review.

Proposed human exposure levels should be reasonably conservative, defensible, and sufficiently protective to avoid a substantial present or potential hazard to people for the forecasted duration of the contamination. A proposed alternate concentration limit that does not exceed an excess lifetime risk of fatal cancer on the order of 10^{-4} is acceptable for an average exposed individual at the point of exposure, when considering the potential for the health risks from human exposure to known or suspected carcinogens contained in untreated ground-water used for drinking water.

Protecting Water Resources

The exposure assessment must identify the point of compliance, where the proposed alternate concentration limit will be measured; and the points of exposure, where the human health and environmental exposures could occur. The assessment identifies the maximum permissible levels of hazardous constituents at the point of compliance that are protective of human health and the environment at the point of exposure. This is accomplished by evaluating human and environmental exposure to each of those constituents evaluated in the hazard assessment, and then showing the proposed alternate concentration limit will not result in an unacceptable exposure of human health or the environment to those hazards. The exposure assessment should include the following:

(1)   The exposure assessment evaluates the pathways the hazardous constituents will likely follow and the concentration or dose those constituents will likely produce at the location where humans or environmental populations could be reasonably exposed. All likely pathways that could transport significant amounts of hazardous constituents in the ground water and hydraulically connected surface water should be identified and evaluated. The hazardous constituent concentrations and projected distributions for each pathway should be best estimates or reasonably conservative representations of the rate, extent, and direction of the constituent transport.

The ground-water pathway evaluation provides projected contaminant distributions, including contaminant transport, degradation, and attenuation mechanisms between the point of compliance and the point of exposure. The evaluation generally provides information on: (a) site hydrogeologic characteristics, including ground-water flow direction and rates; (b) background water quality; and (c) estimated transport rates, geochemical attenuation, and concentrations of hazardous constituents in the ground water and hydraulically connected surface water. Projections should be calibrated on the basis of site-specific information. The projected attenuation rate may rely on constituent concentration measurements at the point of compliance and the point of exposure, taken over an adequate period of time, when there is great uncertainty in the attenuation rate derived from laboratory measurements or literature sources.

(2)   The pathway evaluation provides the spatial distribution of the various hazardous constituents of existing contaminant plumes. This information can be used to calibrate contaminant fate and transport models in the exposure assessment and also identifies the components of the source term that have already been released from the tailings. The contaminant extent characterization includes: (a) the type and distribution of hazardous constituents in the ground water and the source(s) of the contamination; (b) the monitoring program used to delineate and characterize hazardous constituent distribution; and (c) supporting documentation of the sampling, laboratory analysis, and quality assurance programs that show the fulfillment of the site monitoring programs. Such information is used to assess present human and environmental population exposure to elevated concentrations of hazardous constituents, calibrate contaminant transport models, and evaluate projected future exposures. Computer codes may be used to evaluate the pathways for hazardous constituent transport. The acceptance criteria for ground-water fate and transport computer modeling are contained in standard review plan Section 4.4.3.

(3)   The human exposure evaluation considers two potential exposure pathways:
(a) ingestion of contaminated water and (b) ingestion of contaminated foods. Other
pathways that may impact human health, such as dermal contact and inhalation, are
also to be considered, but need not always be assessed, unless it is determined that
these exposures could result in significant hazards to human health or the environment.

Human exposure is evaluated primarily on the basis of the extent to which people are
using, and are likely to use, contaminated water from the site. Site-specific water uses
are determined on the basis of the following considerations: (a) ground-water quality in
the site area and present water uses; (b) statutory or legal constraints and institutional
controls on water use in the site area; (c) federal, state, or other ground-water
classification criteria and guidelines; (d) applicable water-use criteria, standards, and
guidelines; and (e) availability and characteristics of alternate water supplies.

Exposure determinations should consider existing and potential water uses. Potential
uses include those that are reasonably expected to occur (i.e., anticipated use) and
uses that are compatible with the untreated background water quality (i.e., possible
use). Past water uses may be included as existing or potential uses. Water resource
classification of existing and potential water use should include (a) domestic and
municipal drinking water use; (b) fish and wildlife propagation, (c) special ecological
communities uses: and (d) industrial, agricultural, and recreational uses. The
classification of existing and potential water uses at the facility should be consistent with
federal, state, and local water use inventories.

The cumulative effects of human exposure to hazardous constituents at the proposed
alternate concentration limits, and to other constituents present in contaminated ground-
water, will be maintained at a level adequate to protect public health. The combined
effects from both radiological and non-radiological constituents should be considered.
Guidance for cumulative impact assessment is contained in NUREG–1748 (NRC, 2001)
and additional guidance is found in Council on Environmental Quality (1997).

(4)   Potential responses of environmental or non-human populations to the various
hazardous constituents are evaluated if such populations can realistically be exposed to
contaminated ground water or hydraulically connected surface water. Terrestrial and
aquatic wildlife, plants, livestock, and crops are included in this evaluation. A detailed
environmental exposure evaluation should be performed in the absence of available
information that could readily be used to show there will be no substantial environmental
impacts caused by ground-water contamination from the site. The evaluation should
provide: (a) inventories of potentially exposed environmental populations;
(b) recommended tolerance or exposure limits; (c) contaminant interactions and their
cumulative effects on exposed populations; (d) projected responses of environmental
populations that result from exposure to hazardous constituents; and (e) anticipated
changes in populations, independent of the hazardous constituent exposure.
Alternatively, the evaluation may demonstrate that environmental hazards are not
anticipated, because exposure will not occur.

The potential for adverse effects, such as (a) contamination-induced biotic changes;
(b) loss or reduction of unique or critical habitats; and (c) jeopardizing endangered

species, should also be described. Aquatic wildlife effects are evaluated by comparing estimated constituent concentrations with federal and state water quality criteria. Terrestrial wildlife exposure to constituents through direct exposure and food-web interactions should be considered. The NRC must initiate special consultation with the U.S. Fish and Wildlife Service, in accordance with 50 CFR Part 17, if endangered or threatened species occur on the site or could be impacted by site activities. NUREG–1748 (NRC, 2001) should be consulted for initiating this consultation.

Agricultural effects from both direct and indirect exposure pathways, crop impacts, reduced productivity, and bioaccumulation of constituents should be considered. Reasonably conservative estimates of constituent concentrations are compared with federal and state water quality criteria to estimate agricultural effects associated with constituent exposure. Additionally, crop exposures through contaminated soil, shallow ground-water uptake, and irrigation, along with livestock exposure through direct ingestion of contaminated water and indirect exposure through grazing, should be assessed.

(5)   Points of exposure are identified. A point of exposure is any location where people, wildlife, or other species could reasonably be exposed to hazardous constituents from ground water contaminated by uranium mill tailings. For example, the point(s) of exposure may be represented by one or more domestic wells that might withdraw contaminated ground water; or it may be represented by springs, rivers, streams, or lakes into which contaminated ground water might discharge. The point of exposure is used to assess the potential hazard(s) to human health and the environment and effects on the ground-water resource.

An alternate concentration limit for a hazardous constituent is established at the point of compliance. The point of exposure may be situated at some distance from the point of compliance, allowing hazardous constituent concentrations to diminish through dispersion, attenuation, or sorption within the aquifer. As a result, an alternate concentration limit may be set at a concentration that is higher at the point of compliance location than a limit that would be protective of human health and environment, as long as the hazardous constituent will not result in an unacceptable hazard to human health and the environment at the point of exposure. In most cases, the point of exposure is located at the downgradient edge of land that will be transferred to either the federal government or the state for long-term institutional control.

The applicant for an alternate concentration limit should make every reasonable effort to keep the point of exposure at the long-term care site boundary. If this cannot be achieved, a good-faith effort must be made to acquire the land between the license area boundary and the point of exposure, for ultimate transfer to the long-term custodian. If the land cannot be acquired through a good-faith effort, then institutional controls other than ownership by the long-term custodian may be initiated. These institutional controls must be enforceable, durable, and legally defensible; and will be applied in addition to the numerical limits of the proposed alternate concentration limit. This approach must be reviewed as an alternative to the specific regulatory requirements contained in 10 CFR Part 40, Appendix A, Criterion 5B(6).

A distant point of exposure[3] may be justified when human or environmental exposure is effectively impossible. This option could be justified on the basis that extremely rugged terrain cannot be physically accessed or the long-term care custodian would ensure that ground water from the contaminated aquifers between the disposal site and the point of exposure would not be used. In some rare instances, a distant point of exposure could be established without invoking land ownership by a long-term custodian. Under these circumstances, the previously described institutional controls should be invoked. Human and environmental exposure are considered effectively impossible when the ground water is inaccessible or unsuitable for use. Land ownership or long-term custody will not be an issue for establishing a distant point of exposure if human and environmental exposure are effectively impossible.

When a distant point of exposure is involved, the applicant must coordinate the use of this option with the NRC. The NRC and the applicant must verify whether the state or the federal government will be the long-term site custodian, after the license is terminated. The applicant must then secure a commitment from that party to take custody of the site. The applicant or the NRC must then secure written assurance that the appropriate federal or state agency will accept the transfer of the specific property, including land in excess of that needed for tailings disposal. Alternate concentration limits may not be established at sites involving a distant point of exposure until the licensee agrees to transfer the title to the land, and the appropriate federal or state government commits to take such land, including the land between the point of compliance and point of exposure that is in excess of the land used for disposal of byproduct material.

If the licensee chooses to keep the mill property under a specific license and apply for an alternate concentration limit as part of a compliance monitoring program, the licensee must still coordinate the use of a distant point of exposure with the NRC as described above.

(6)   The likelihood of human and environmental exposure is determined. The probability of human and environmental exposure is often difficult to establish quantitatively. Consequently, defensible qualitative estimates of the exposure likelihood are often necessary. These can be characterized as either:

(a)   Reasonably likely—when exposure has or could have occurred in the past, or available information indicates that exposure to contamination may reasonably occur during the contamination period.

(b)   Reasonably unlikely—when exposure could have occurred in the past, but will probably not occur in the future, either because initial incentives for water use have been removed, or because available information indicates that no

---

[3]A distant point of exposure refers to a point of exposure that is spatially beyond the area that the appropriate federal or State agency is required to accept for perpetual care under the land transfer provisions of the Uranium Mill Tailings Radiation Control Act of 1978, as amended.

incentives for water use are currently identifiable, based on foreseeable technological developments.

(7)   Exposure impacts are adequately evaluated through time. It is acceptable to project impacts at the point of exposure during a 1,000-year time frame. This is consistent with the design standard of 10 CFR Part 40, Appendix A, Criterion 6(1).

4.3.3.3 Corrective Action Assessment

The applicant's assessment of ground-water corrective action alternatives should be reviewed in conjunction with the hazard assessment and the exposure assessment. Past, current, and proposed practicable corrective actions are identified and evaluated against the costs and benefits associated with implementing each corrective action alternative. The corrective action assessment should demonstrate that the proposed alternate concentration limit is as low as is reasonably achievable, considering practicable corrective actions, as required by 10 CFR Part 40 Appendix A, Criterion 5B(6). A principal way of demonstrating this is by estimating and comparing the benefits imparted by a corrective action measure against the cost of implementing that measure.

For some sites, a corrective action assessment may have already been completed, as part of a ground-water corrective action program under Criterion 5D of Appendix A to 10 CFR Part 40, as described in standard review plan Section 4.4.3. A ground-water corrective action assessment typically (a) identifies several practicable corrective action alternatives; (b) assesses the technical feasibility, costs, and benefits of each alternative; and (c) selects an appropriate corrective action for achieving compliance with the ground-water protection standards established at the site. The corrective action assessment should include the following:

(1)   A complete range of realistic and reasonable corrective action alternatives for achieving compliance with the ground-water standards currently in the license and the proposed alternate concentration limit is described and evaluated. The identified alternatives should be comprehensive, including all engineering-feasible alternatives, both passive and active, or any appropriate sequential combination of alternatives. The analyzed corrective action alternative should not simply be a compendium of the most elaborate and expensive alternatives. The description of each alternative should be conceptual in nature, but contain sufficient detail so the reviewer can independently verify the reasonableness of each corrective action measure. Although conceptual, the alternate descriptions should also contain sufficient detail for completing a coarse cost estimate of each alternative for the cost and benefit analysis.

For past and current corrective actions, site-specific operational and monitoring data should be included to show the effectiveness of those measures. The evaluation may include information from literature sources or documented experience from other sites for those corrective actions that have not been implemented at the site but appear to be practicable. The evaluation should also include projections of the hazardous constituent concentration that each corrective action would likely produce at specific times at the point of compliance and the point of exposure. It is important that the reviewer assure that the range of reasonable corrective actions listed in the application is complete. The suitability of a corrective action should be determined strictly on the technical and engineering information needed to design and implement a particular measure. The economic constraints for implementing a particular measure should not be used to eliminate a corrective action method from the evaluation.

(2)   The direct benefits of implementing the corrective actions have been determined by estimating the current and projected resource value of the pre-contaminated ground water. Estimates of pre-contaminated ground-water value should be based on water rights, availability of alternate water supplies, and forecasted water use demands. The value of a contaminated water resource is generally equal to the cost of a domestic or municipal drinking water supply or the cost of water supplied from an alternate source to replace the contaminated resource. The absence of available alternate water supplies increases the relative value of a potentially contaminated water resource. The indirect benefits are determined by assessing the avoidance of adverse health effects from exposure to contaminated water, the prevention of land value depreciation, and any benefits accrued from performing the corrective action, including timeliness of remediation. The reviewer should verify the water yields; costs for developing alternate water supply sources; and legal, statutory, or other administrative constraints on the use and development of the water resources.

(3)   The costs associated with performing a corrective action alternative to achieve the target concentrations include (a) the capital costs for designing, and constructing the alternative; (b) operation and maintenance costs; (c) costs associated with demonstrating compliance with the standards; and (d) decommissioning costs after the corrective action is completed.

(4)   The "as low as is reasonably achievable" analysis is performed on target concentration levels that are at or below the limit determined to be protective of human health and the environment. At least three target concentration levels that can reasonably be attained by the practicable corrective actions should be evaluated. The goals should be (a) meaningfully different, (b) reasonably attainable by practicable corrective action, and (c) at or below the levels identified in the hazard assessment.

The "as low as is reasonably achievable" analysis typically considers (a) the direct and indirect benefits of implementing each corrective action to achieve the target concentration levels; (b) the costs of performing the corrective action to achieve the target concentrations; and (c) a determination whether any of the evaluated corrective action alternatives will reduce contaminant levels below the proposed alternate concentration limit, considering the benefits and costs of implementing the alternative.

The applicant should also provide a comparison among the costs associated with performing the various corrective action alternatives to achieve the target concentrations, the value of the pre-contaminated ground-water resource, and the benefits of achieving each target concentration. A proposed alternate concentration limit is considered as low as is reasonably achievable if the comparison of the costs to achieve the target concentrations lower than the alternate concentration limit are far in excess of the value of the resource and the benefits associated with performing the corrective action alternative. If the value and benefits clearly exceed the costs or the comparison is nearly equal, the proposed alternate concentration limit should be revised

Protecting Water Resources

to the lower target concentration providing the greatest value and benefit compared to the cost.

The cost and benefit analysis should not be limited to a simple financial accounting of the costs for each corrective action alternative. Costs and benefits should also be discussed for qualitative subjects, such as environmental degradation or enhancement. The cost and benefit analysis is not simply a mathematical formula from which to justify economic parameters. Other qualitative factors should be discussed and weighed in the decision. The cost and benefits analysis provides input to determine the relative merits of various corrective action alternatives; however, the proposed alternate concentration limit must ultimately assure protection of public health and the environment.

The as low as is reasonably achievable analysis for non-radiological constituents should be similar to the as low as is reasonably achievable analysis for radiological constituents except a "dollar per person-rem avoided" value would not be calculated. Additionally, once nonradiological constituent are below regulatory maximum concentration levels, the licensee ha no further obligation to reduce the constituent concentrations.

4.3.3.4Examination of the Compliance Monitoring Program

Standard review plan Section 4.4.3 provides the acceptance criteria for corrective action assessments, corrective action monitoring, and compliance monitoring. The reviewer should examine the existing compliance monitoring program at a licensed mill tailings facility, if a proposed alternate concentration limit is found acceptable.

Specifically, the compliance monitoring program should monitor all ground-water exposure pathways to assure that any potential exceedances of the proposed alternate concentration limit will be detected before the license is terminated. The compliance monitoring well locations should not be restricted solely to the point of compliance. Some locations between the point of compliance and the points of exposure should be included to assure the identified aquifer attenuation mechanisms are reducing the hazardous constituent concentrations to the predicted levels. The applicable maximum contaminant level, background concentration, or other maximum permissible limit should be used as the compliance monitoring limit for wells at the points of exposure, in those cases where compliance monitoring is conducted at the points of exposure.

4.3.4 Evaluation Findings

The following conclusions may be presented in the technical evaluation report, if the staff review, as described in standard review plan Section 4.3 results in the acceptance of the hazard assessment, exposure assessment, and corrective action assessment supporting the proposed alternate concentration limit:

The staff completed its review of the proposed alternate concentration limit for ground-water compliance at the _____ uranium mill tailings facility. This review included an evaluation using the review procedures in Section 4.3.2 and the acceptance criteria outlined in Section 4.3.3 of this standard review plan.

The licensee conducted an acceptable hazard assessment by considering present and potential human health and environmental hazards, including human cancer risk from exposure to radioactive and non-radioactive constituents and other health hazards resulting from the chemical toxicity of the constituents. The source term for constituents of concern and the extent of ground-water contamination have been acceptably characterized.

The licensee conducted an acceptable exposure assessment. The point of exposure has been identified and is acceptably sited at the downgradient edge of the affected land. When a distant point of exposure is used, written assurance has been secured, either by the licensee or NRC, that the appropriate federal or state agency will accept the transfer of the specific property, including land in excess of that needed for tailings disposal. The transport of the hazardous constituent in ground water and surface water has been defined and any adverse effects on water quality, including present and future impacts have been assessed.

The human cancer risk and other health and environmental hazards from exposures to hazardous constituents have been evaluated and are acceptable, including (a) identification of maximum levels permissible at the point of compliance; (b) evaluation of health and environmental hazards using water classification and use standards and existing and anticipated water uses; (c) appropriate consideration of impact, based on site-specific water uses; (d) consideration of ingestion of contaminated water and food; (e) consideration of response of environmental and non-human populations to the various hazardous constituents, including terrestrial and aquatic wildlife, plants, livestock, and crops; and (f) consideration of potential damage to physical structures.

The acceptable corrective action assessment includes (1) an assessment of ground-water corrective actions dealing with identification of practicable corrective action alternatives; (2) an evaluation of ability of corrective action to reduce contaminant levels appropriately; (3) a demonstration that action will achieve desired concentration levels; and (4) demonstration that practicable corrective actions are not likely to result in reduction of contamination below the proposed alternate concentration limit, and that alternate concentration limits are, therefore, as low as is reasonably achievable.

The NRC staff concludes that the information submitted to support the proposed alternate concentration limit(s) at the _____ uranium milling facility is acceptable and complies with the following criteria in 10 CFR Part 40, Appendix A, Criterion 5B, which requires NRC to establish a list of hazardous constituents, concentration limits, a point of compliance, and a compliance period; Criterion 5C, which contains a table of secondary concentration limits for certain constituents when they are present in ground water above background concentrations; Criterion 5F, which requires that where ground-water impacts from seepage are occurring at an existing site, action must be taken to alleviate the conditions that lead to seepage, and ground-water quality must be restored, including technical specifications for the seepage control system and implementation of a quality assurance program; Criterion 5G, which requires licensees/operators to perform site characterization; Criterion 6(1), which provides performance lifetime and radioactive material release standards; and Criterion 7A, which establishes detection, compliance and corrective action monitoring programs in support of a tailings disposal system proposal. The information also complies with 10 CFR 40.31(f), which requires inclusion of an environmental report in the license application, and

Protecting Water Resources

10 CFR 51.45, which requires a description of the affected environment containing sufficient data to aid the Commission in its conduct of an independent analysis.

4.3.5 References

Council on Environmental Quality. "Considering Cumulative Effects Under the National Environmental Policy Act." Washington, DC: Council on Environmental Quality. 1997.

NRC. NUREG–1748, "Environmental Review Guidance for Licensing Actions Associated with NMSS Programs." Washington, DC: NRC. 2001.

4.4   Ground-Water Corrective Action and Compliance Monitoring Plan

The staff should review any ground-water corrective action and compliance monitoring plans that may be presented by the licensee either as a part of the reclamation plan, or as a separate licensing submittal. A separately submitted corrective action and compliance monitoring plans will contain much of the same information that is required for the reclamation plan (e.g., a site characterization plan). Any information that was presented in a previously approved reclamation plan may be incorporated by reference. For review of some information, the reviewer may use review procedures in other chapters of this standard review plan.

4.4.1 Areas of Review

In determining compliance, the reviewers should consider the information specified in Criteria 1–8 of Appendix A of 10 CFR Part 40 that is relevant to the technical adequacy of the ground-water corrective action and compliance monitoring plans. Models of unsaturated flow and transport can be used if the tailings pile is located in the unsaturated zone. A reactive transport model of the plume of hazardous constituents for the saturated zone away from the mill tailings pile should be constructed if the licensee takes credit for chemical processes that may mitigate the spread of contaminants. The technical adequacy of any detailed models should be reviewed. Findings from detailed models (that incorporate complexities not treated in any large-scale numerical models) can be used as input to a large-scale numerical model of ground-water flow and transport for the site. Models should be calibrated using site data.

The staff should review the following aspects of ground-water corrective action and compliance monitoring plans.

(1)   The sufficiency of data and parameters.

(2)   The technical bases for parameter ranges.

(3)   Descriptions of features and physical phenomena.

(4)   Use of alternate models.

(5)   Consistency of models.

(6)   Waste management practices.

(7)   Site access controls.

(8)   Ground-water monitoring plans.

(9)   Design, operation, and inspection of surface impoundments.

(10)   Surety.

4.4.2 Review Procedures

In conducting the review of the technical adequacy of the ground-water corrective action and compliance monitoring plans, the staff should recognize that review procedures and models used in the technical assessment of the selected ground-water cleanup methods, cleanup time, and sureties may range from detailed, small-scale process models to large-scale, simplified models. The small-scale process models incorporate the important complexities and mechanisms that govern the evolution of the hazardous constituent plume, while the large-scale simplified models do not consider all the important complexities. Model adequacy should be evaluated regardless of the level of complexity. The following areas should be evaluated:

(1)   The staff should evaluate the sufficiency of the data and parameters supporting models considered in any site-scale numerical model used to estimate the cleanup time. The staff should also evaluate the technical basis for data on design features, physical phenomena, geology, hydrology, geotechnical engineering, and geochemistry used to model or assess ground-water cleanup. This basis may include a combination of techniques such as laboratory experiments and site-specific field measurements.

        The reviewer should evaluate whether additional data are likely to provide new information that could invalidate the modeling results and significantly affect the corrective action and compliance monitoring plans.

(2)   The reviewer should evaluate the technical bases for parameter ranges, probability distributions, or bounding values. The reviewer should determine whether the parameter values are derived from site-specific data or, alternatively, an analysis is included to show that the assumed parameter values lead to a conservative assessment of performance.

        The staff should examine the initial conditions and boundary conditions used in sensitivity analyses for consistency with available data. The staff should also consider the temporal and spatial variations in boundary conditions and source terms used to support the ground-water corrective action and compliance monitoring plan.

        The staff should evaluate the licensee assessment of uncertainty and variability in parameters used in the modeling. The reviewer should determine whether uncertainty in data due to both temporal and spatial variations in conditions affecting the ground-water cleanup and estimation of cleanup time was incorporated into parameter ranges.

(3)   The staff should examine the descriptions of features and physical phenomena and the descriptions of the geological, hydrological, geotechnical, and geochemical aspects of the mill tailings and the underlying aquifer. The staff should verify that the descriptions are adequate and that the conditions and assumptions used in the modeling are realistic or reasonably conservative and are supported by the body of data presented in the descriptions. The staff should assess the technical bases for these descriptions and for incorporating them in the numerical model of the site.

The reviewer should examine the technical bases for the identification of hazardous constituents from the mill tailings that have entered the underlying aquifer and surface water bodies. The staff should evaluate how these constituents have been incorporated into any detailed models. The staff should also verify that, given the concentrations and locations of the hazardous constituents, estimates of cleanup time and surety amounts are reasonable.

The reviewer should examine the assumptions used to develop any model of reactive transport that accounts for site geochemical processes. These processes may include phase changes induced by interaction of contaminants with ground water and surface water. The modeling should consider available data about the native ground-water downgradient of the tailings pile, the geochemical environment, hydraulic and transport properties, and the spatial variations of properties of aquifers and ground-water volumetric fluxes along the flow paths.

The staff should evaluate the initial and boundary conditions and how they have been propagated through the models. For example, the reviewer should determine whether the conditions and assumptions used in the site-scale model are consistent with other conditions and assumptions used in any model that describes the flow and transport of hazardous constituents from the mill tailings.

(4)   The staff should evaluate models used for the ground-water cleanup and estimation of cleanup time. The staff should examine the model parameters in the context of available site characterization data, laboratory experiments, and field measurements. Where appropriate, and when surety estimates are highly uncertain, the reviewer should use an alternate site model to evaluate the effects on the technical assessment of ground-water cleanup and estimation of cleanup time.

(5)   The staff should evaluate the output from any site model of ground-water cleanup and the estimation of cleanup time and compare the results with an appropriate combination of site characterization and design data.

The staff should examine the model results obtained by the licensee against comparable mathematical models to judge their robustness. The reviewer should use an alternate model to evaluate selected parts of the licensee model results, as appropriate. The reviewer should evaluate whether the licensee has appropriately reduced the dimensionality and complexity of models. The dimensionality of models, heterogeneity of aquifer parameters, and significant process couplings may be reduced if it is shown that the simplified model bounds the prediction of the more complex model.

The staff should evaluate the acceptability of the sensitivity analyses used to support the model of the ground-water cleanup and the estimation of cleanup time.

(6)   The staff should verify that waste management practices are in compliance with environmental protection regulations.

(7)   The reviewer should assess whether site access controls during the cleanup period are sufficient to prevent significant hazards to human health and the environment.

(8)   The staff should evaluate whether the ground-water monitoring system is sufficient to verify the performance of the selected cleanup strategy, and to monitor the long-term performance of any on-site tailings disposal cells.

(9)   The staff should ensure that any surface impoundments constructed as part of the program are designed to meet the requirements of 10 CFR Part 40, Appendix A criteria and are included in the dam safety program, if appropriate. The reviewer should also verify that adequate inspection, documentation, and reporting procedures exist for tailings or waste retention systems.

(10)   The staff should confirm that the applicant has provided adequate financial surety. This confirmation may be conducted using cost estimating software such as the RACER 2000™ computer code (Talisman Partners, Ltd., 2000). Guidance on the preparation of sureties and cost estimates is available in Appendix C of this standard review plan and in NRC (1988, 1997).

4.4.3 Acceptance Criteria

In 10 CFR Part 40, Appendix A, Criterion 5D, NRC requires tha,t if the ground-water protection standards established under Criterion 5B(1) of 10 CFR Part 40, Appendix A are exceeded at a licensed site, a corrective action program must be put into operation as soon as is practicable, and in no event later than 18 months after the Commission finds that the standards have been exceeded. Unless otherwise directed by the Commission, before putting the program into operation, the licensee should submit the supporting rationale for the proposed corrective action program. The objective of the program is to return hazardous constituent concentration levels in ground-water to the concentration limits set as standards. The licensee should provide an assessment of practicable corrective actions available for returning contaminant concentrations to the standards established in the license. The corrective action assessment incorporates information and findings from the site characterization activities, which are described in standard review plan Section 4.1.3. Site specific characteristics may have a strong influence on which corrective action alternative will be practicable for a particular site. If additional site characterization is needed, details of the characterization plan should be included.

The corrective action should result in conformance with the established concentration limits, address either removing the hazardous constituents or treating them in place, and should include a program to monitor compliance with cleanup standards. Regulations do not require any specific designs or methods to be used for the ground-water corrective action program. Because of the nearly limitless possibilities for designing and implementing ground-water corrective actions, staff reviewers should focus on the technical feasibility from an engineering

Protecting Water Resources

perspective and evaluate whether the proposed design is likely to result in timely compliance with established concentration limits and whether the monitoring program is adequate to verify the effectiveness of the design. Useful guidance for the application of ground-water flow and transport modeling can be obtained from American Standard for Testing and Material D 5447, D 5490, D 5609, D 5611, D 5718, D 5880, and D 5981.

The ground-water corrective action and compliance monitoring plans are acceptable if they meet the following criteria.

(1)   Sufficient data are available to adequately define relevant parameters and to support models, assumptions, and boundary conditions necessary for developing detailed and site-scale models of the ground-water cleanup and the estimation of cleanup time. The data are also sufficient to assess the degree to which processes related to the ground-water cleanup that affect compliance with the technical criteria in Appendix A of 10 CFR Part 40 have been characterized. Information required for site-scale reactive transport models can include:

    (a)   Site description.

        (i)   Chronology/history of uranium milling operations.

        (ii)   List of known leaching solutions and other chemicals used in the milling process.

        (iii)   Summary of known impacts of the site activities on the hydrologic system and background water quality.

        (iv)   Quantity and chemical/textural characteristics of wastes generated at the mill site.

        (v)   Information pertaining to surrounding land and water uses.

        (vi)   Meteorological data for the region including precipitation and other data to support estimates of evapotranspiration.

    (b)   Description of hydrogeologic units.

        (i)   Hydrostratigraphic cross sections/maps.

        (ii)   Hydrogeologic units that constitute the aquifer(s).

        (iii)   Description of perched aquifers (areal/volumetric extent).

        (iv)   Description of the unsaturated zone (thickness, extent).

        (v)   Geologic characteristics (presence of layers, continuity, faults).

    (c)   Data on the hydraulic and transport properties of each aquifer.

    (i)   Hydraulic conductivity.

    (ii)   Thickness of each unit.

    (iii)   Hydraulic head contour maps (of each aquifer).

    (iv)   Information on background horizontal and vertical hydraulic gradients and temporal variations to determine flow directions.

    (v)   Vertical hydraulic gradients and inter-aquifer flow within and between multiple aquifer systems.

    (vi)   Effective porosity

    (vii)   Storativity or specific yield (for transient simulations).

    (viii) Longitudinal, vertical and horizontal transverse dispersivity.

    (ix)   Retardation factors.

(c)   Data on regional recharge rates and ground-water/surface-water interactions with nearby streams, rivers, or lakes.

    (i)   Areal recharge rates.

    (ii)   Information on water fluxes to and from rivers, aquifers, and surface water bodies.

    (iii)   Data on surface water bodies (e.g., stream flow rates, dimensions of nearby surface water bodies).

    (iv)   Concentration of hazardous constituents in surface water bodies

(d)   Characteristics of the mill tailings.

    (i)   Identification of contaminant source terms.

    (ii)   Hydraulic properties of mill tailings material.

    (iii)   Unsaturated flow and transport parameters of mill tailings material.

    (iv)   Design and materials for mill tailings cover.

    (v)   Information on the spatial and temporal distribution of seepage fluxes from the mill tailings to the upper-most aquifer (including the historical variation in rates).

    (vi)   Information on mill tailings draining mechanisms and drainage volume.

(vii)   Geotechnical properties of the mill tailings and their temporal variation due to drainage of leachates

(viii)   Tailings volume.

(ix)   Data on the volume, chemical and mineralogical characteristics, and concentration of mill tailings and tailings solution/leachate.

(x)   Mass of hazardous constituents placed in the tailings pile and other disposal or storage areas.

(e)   Data on geochemical conditions and water quality.

(i)   Concentration of hazardous constituents.

(ii)   Background (baseline) ground-water quality.

(iii)   Delineation of the nature and extent of the hazardous constituent plume.

(iv)   Characterization of subsurface geochemical properties.

(v)   Identification of attenuation mechanisms and estimation of attenuation rates.

(vi)   Mass of hazardous constituents in the aquifer.

(f)   Site cleanup data.

(i)   Information on grout curtains, slurry walls, drains, interceptor ditches, and other facilities designed to reduce the spreading of the hazardous constituent plume (if used).

(ii)   Information on pumping, injection, and sampling wells (coordinates, depths, completion diagrams, flow rates).

(iii)   Pumping/injection rates and rate history for each well (if pumping has been ongoing).

(iv)   Information on the presence or the absence of liners for the mill tailings pile and evaporation ponds.

(v)   Mass of hazardous constituents recovered to date.

Sufficient data are available to justify models used to validate the ground-water corrective action plan. American Standard for Testing and Materials D 5490 provides acceptable guidance for comparing model simulations to site-specific information. Alternatively, in the case of sparse data and/or low confidence in the quality of available data or data interpretations, the licensee demonstrates by sensitivity analyses or other methods that the proposed ground-water corrective action plan is appropriate, and the contingency built into the surety is compatible with the uncertainties. American Standard for Testing and Materials D 5611 provides acceptable guidance for conducting sensitivity analyses on ground-water flow models. Guidance on preparing cost estimates and establishing sureties for uranium mills is provided in the "Technical Position on Financial Assurances and Reclamation, Decommissioning, and Long-Term Surveillance and Control of Uranium Recovery Facilities" (NRC, 1988).

Sufficient information is provided to substantiate that any mathematical flow and transport modeling approach is appropriate for site conditions considering (i) factors pertaining to the specific purpose or intended use of the model(s); (ii) the flow media at the site and along the flow path from the mill tailings to the point of compliance, and downgradient to it, including aquifer properties and transport parameters (e.g., porous media versus fracture flow, aquifer confinement, the number of active layers); (iii) modeling assumptions (e.g., steady-state versus transient flow, assignment of initial and boundary conditions); and (iv) model-related factors (e.g., underlying flow equations; solution methods; model history; model verification, validation and calibration; expertise and experience of the personnel responsible for model development; and quality of model documentation). Amiercan Standard for Testing and Materials D 5718 provides guidance for documenting ground-water flow model applications.

An adequate assessment is provided of the low and high permeability features (heterogeneities), their spatial distributions, and statistical properties; and the available and acquired data are suitable and sufficient for modeling based on observations, independent analyses, or published reports and databases of those features.

Initial and boundary conditions used by the licensee in modeling the ground-water cleanup are justified by the available data, are used consistently throughout the modeling process, and are adequately documented. American Standard for Testing and Materials D 5609 provides acceptable guidance for defining boundary conditions for ground-water flow models.

Where sufficient data do not exist, the definition of parameter values and conceptual models are based on appropriate sources from the literature or are otherwise technically justified.

Adequate site geochemical data are provided. Contaminants are identified sufficiently to support the ground-water corrective action plan and models. In addition to helping set cleanup goals, background water chemical data support assessments of geochemical evolution as ambient ground water is restored in the subsurface. Generally, a three-dimensional delineation of contaminant distribution and a source term are necessary for defining needed actions and for model development. The important geochemical parameters that should be delineated for both contaminated and background waters are pH, Eh, dissolved oxygen, temperature, major cation and anion concentrations, and concentrations of potential contaminants. Host rock properties affect both the water chemistry and the specific geochemical mechanisms affecting contaminants. Identifying possible attenuation mechanisms ensures that cleanup is based on reasonable models for contaminant transport.

Protecting Water Resources

(2)   Parameter values, assumed ranges, probability distributions, and/or bounding
       assumptions used in the modeling of ground-water cleanup are technically defensible
       and reasonably account for uncertainties and variabilities. The technical bases for each
       parameter value, ranges of values, or probability distributions used in the modeling of
       the ground-water cleanup are provided.

       Sensitivity analyses are provided that (i) identify aquifer flow and transport parameters
       that are expected to significantly affect the site model outcome; (ii) test the degree to
       which the performance of the ground-water cleanup may be affected if a range of
       parameter values must be used as input to the model due to sparsity of, or uncertainty
       in, available data; and (iii) test for the need for additional data.

       Sufficient bases are provided for parameter values, representative parameter values are
       taken from the literature, and the bounds and statistical distributions are provided for
       hydrologic and transport parameters that are important to the estimation of cleanup time
       and that are included in the modeling of the ground-water cleanup.

       Site data fitted to theoretical models compare reasonably well. American Standard for
       Testing and Materials D 5490 provides guidance for comparing ground-water flow model
       simulations to site-specific information. If there is departure of site data from the
       theoretical model, then an alternate model is considered. The assumptions used in
       modeling are consistent with site data and observations.

       Models used to describe local phenomena, such as the fluxes through the tailings pile,
       are based on consistently applied conditions.

(3)   Important design features, physical phenomena, and consistent and appropriate
       assumptions are identified and described sufficiently for incorporation into any modeling
       that supports the ground-water cleanup, including the estimate of cleanup time, and the
       technical bases are provided. Detailed models and site-scale models used to support
       the corrective action plan, or other supporting documents, and identify and describe
       aspects that are important to the cleanup and the estimate of cleanup time.

       The licensee delineates the extent of the hazardous constituent plume, contaminant flow
       paths in the aquifer considering natural site conditions, any effects that can be expected
       to result from construction of additional facilities and operations (i.e., tailings ponds,
       evaporation ponds, excavations), and events that may affect the spatial and temporal
       distribution of the hazardous constituent plume. More specifically, the licensee's models
       of the ground-water cleanup consider and are consistent with (i) natural climatic,
       geologic, and hydrologic conditions at the site and in the vicinity of the site; (ii) tailings
       pile design and construction features and their potential impact on local recharge and
       consequent flow paths in the aquifer; (iii) geochemical and other processes that can
       affect the performance of the ground-water cleanup and estimation of cleanup time; and
       (iv) future events, including additional construction and changes of plans for operations
       that may occur at the site. The licensee also has determined the range of
       concentrations of hazardous constituents that can be expected in the aquifer and their
       changes with time during the ground-water cleanup.

The licensee estimates the total mass of hazardous constituents produced by the leaching process and the quantity of the mass that is in the mill tailings, the aquifer, in surface water bodies (including evaporation ponds, disposal cells, nearby ponds, and rivers) and the portion that has been removed by means of the ground-water cleanup, and accounts for the mass that will be removed for final disposal.

The licensee makes reasonable assumptions, if taking credit for dispersion of hazardous constituents and consequent reduction of concentrations during transport from the mill tailings to the point of compliance, for such processes as mechanical dispersion and mixing with native ground water and surface water. These assumptions are based on available data about the hydraulic and transport properties of the site and the spatial variations of properties of aquifers and ground-water volumetric fluxes along the flow paths.

The licensee provides an adequate basis for considering the effect of any reactive transport and geochemical processes in simulating the ground-water cleanup operation, if taking credit for sorption or any other geochemical reaction of hazardous constituents and consequent reduction or retardation of concentrations during transport from the mill tailings. Predicting the effects of proposed ground-water cleanup actions may include forward, site-specific contaminant transport modeling. Often, such modeling has taken a simple approach employing a retardation factor to describe all geochemical effects on contaminant concentration. This approach may be too simplistic. The use of a constant retardation factor and the neglect of speciation and water-mineral reactions is likely to lead to prediction errors. Reactive transport models using codes such as PHREEQC Version 2 (Parkhurst and Appello, 1999) are acceptable for constructing a geochemical model for the site. Hostetler and Erickson (1993) discuss examples of the effect of extending reactive transport models beyond simply including retardation in advective-dispersive models. In one example involving cadmium transport at a uranium mill tailings site, concentration profiles from the site suggest the importance of otavite ($CdCO_3$) solubility control on aqueous cadmium in the low-pH zones near the tailings pond, and the inadequacy of modeling sorption alone.

Reactive transport models incorporate thermodynamic data on solid phases and aqueous species, allowing the mass action calculations that determine estimated aqueous concentrations and solid phase evolution. Thermodynamic parameters constitute a major source of uncertainty in geochemical modeling [see Murphy and Shock (1999) for a discussion of uranium], with potentially large effects on predicted aqueous ion concentrations. Therefore, geochemical modeling supporting ground-water corrective action plans includes sensitivity analyses that provide assurance that contaminant concentrations will not be underestimated. Likewise, any kinetic models employed are subjected to critical analysis because of the large influence of kinetic effects at low temperatures.

Reactive transport model results are subject to the assumptions and limitations of the conceptual and numerical models employed. For example, Zhu et al.[4] list model limitations and briefly discuss how they may affect predictions. Geochemical limitations include:

(a)   The assumption of local equilibrium (i.e., kinetic rates were not employed).

(b)   Modeled porosity not being affected by reactions affecting the solid phase.

(c)   Omitting colloidal transport.

(d)   Neglecting density effects due to varying total dissolved solids.

(e)   Simplifying the mineralogical suite.

(f)   Neglecting surface reactions such as ion exchange.

(g)   Relying on bulk mineralogy rather than on mineral surface compositions.

Limitations such as these are typically due to factors such as lack of data, inadequate computational equipment, or insufficient model development. Consideration of model limitations and their effects on uncertainty is an important component of the review by the NRC.

The numerical model of the site constructed by the licensee incorporates site-specific information, is adequately validated and calibrated, and reasonably represents the physical system. American Standard for Testing and Materials Reports D 5490 and D 5981 provide guidance for ground-water flow model validation and calibration. The professional experience and judgment of the reviewer should be applied in assessing these aspects of the analyses.

The licensee identifies and properly integrates factors that are expected to affect, or that are affected by, the ground-water cleanup. These include, but are not limited to, the spatial and temporal variation of the flux of leachates from the mill tailings to the underlying aquifer, drainage mechanisms of leachates from the mill tailings, spatial variability in flow and transport properties of the aquifer underlying the mill tailings, and geochemical processes that may affect the concentrations of hazardous constituents.

The licensee evaluates and documents the degree of conservatism in modeling the ground-water cleanup, and the level of conservatism presumed by the licensee is commensurate with the data and conceptual model uncertainty.

---

[4]Zhu C., F.Q. Hu, and D.S. Burden. "Multi-Component Reactive Transport Modeling of Natural Attenuation of an Acid Ground-Water Plume at a Uranium Mill Tailings Site." *Journal of Contaminant Hydrology.* 2001. Accepted for publication.

Protecting Water Resources

(4)   Alternate modeling approaches consistent with available data and current scientific
        understanding are investigated where necessary, and results and limitations are
        appropriately factored into the ground-water corrective action plan. The licensee
        provides sufficient evidence that relevant site features have been considered, that the
        models are consistent with available data and current scientific understanding, and that
        the effects on cleanup time have been evaluated. Specifically, the licensee adequately
        considers alternate modeling approaches where necessary to incorporate uncertainties
        in site parameters and ensure they are propagated through the modeling.

        Uncertainty in data interpretations is considered by analyzing reasonable conceptual
        models that are supported by site data, or by demonstrating through sensitivity studies
        that the uncertainties have little impact on the ground-water corrective action plan.

(5)   The site-scale model for ground-water cleanup provides results consistent with the
        output of detailed or site data. Specifically, the site model is consistent with detailed
        models of geological, hydrological, and geochemical processes for the site. For
        example, for flow and transport through the aquifer, hydraulic conductivity distributions
        are reasonably consistent with sensitivity studies of the range of hydraulic
        conductivities and varying statistical distributions, field observations, and laboratory
        tests, when applicable.

        The licensee documents how the model output is validated in relation to
        site characteristics.

        Where appropriate, in developing the site model for ground-water cleanup, the licensee
        considers and evaluates alternate models that are reasonably justified by the available
        database, with reasonable values assigned to distribution statistics to compensate for
        limited data availability.

        The licensee uses numerical and analytical modeling approaches reflecting varying
        degrees of complexity consistent with information obtained from site characterization.

        The licensee employs the upper and lower bounds of input parameter ranges to
        examine the robustness of the modeling.

(6)   Adequate waste management practices are defined.

        The disposition of effluent generated during active remediation is addressed in the
        corrective action plan. Appendix F to this standard review plan contains NRC staff
        policy for effluent disposal at licensed uranium recovery facilities for conventional mills.
        When retention systems such as evaporation ponds are used, design considerations
        from erosion protection and stability along with construction plans reviewed by a
        qualified engineer are included. Evaporation and retention ponds should meet the
        design requirements of 10 CFR Part 40, Appendix A, Criterion 5A. Ideally, the ponds
        should have leak detection systems capable of reliably detecting a leak from the pond
        into the ground water and should be located where they will not impede the timely
        surface reclamation of the tailings impoundment.

Protecting Water Resources

If water is to be treated and reinjected, either into an upper aquifer or into a deep disposal well, the injection program is approved by the appropriate State or Federal authority. For release of this waste to surface waters, existing licensees must meet the requirements of 10 CFR 20.1302(b)(2), and should demonstrate that doses are maintained as low as reasonably achievable (ALARA). NRC has no specific requirements for non-radiological constituents, and may adopt the appropriate State limits. Anticipated discharge must be described in enough detail to evaluate environmental impacts. Appropriate State and Federal agency permits should be obtained in accordance with 10 CFR 20.2007.

(7)   Appropriate site access control is provided by the licensee.

Site access control should be provided by the licensee until site closure to protect human health and the environment from potential harm. Site access is controlled by limiting access to the site with a fence and by conducting periodic inspections of the site.

(8)   Effective corrective action and compliance monitoring programs are provided.

Licensees are required, by Criterion 7 of Appendix A to 10 CFR Part 40, to implement corrective action and compliance monitoring programs. The licensee monitoring programs are adequate to evaluate the effectiveness of ground-water cleanup and control activities, and to monitor compliance with ground-water cleanup standards. The description of the monitoring program includes or references the following information:

(a)   Quality assurance procedures used for collecting, handling, and analyzing ground-water samples.

(b)   The number of monitor wells and their locations.

(c)   A list of constituents that are sampled and the monitoring frequency for each monitored constituent.

(d)   Action levels that trigger implementation of enhanced monitoring or revisions to cleanup activities (i.e., timeliness and effectiveness of the corrective action).

For corrective action monitoring:

The same wells used to determine the nature and extent of contamination may be used to monitor the progress of ground-water corrective action activities. However, once the extent of contamination is delineated, it may be possible to adequately monitor compliance with fewer wells. Once selected, major changes to monitored locations are avoided, because it is important to be able to directly compare measurements made at different times.

Licensees choose a monitoring interval that is appropriate for monitoring corrective action progress. Not all hazardous constituents need to be monitored at each interval. It is generally acceptable for licensees to choose a list of more easily measured constituents that serve as good indicators of performance.

These indicators include conservative constituents that are less likely to be attenuated, such as chloride, total dissolved solids, and alkalinity. However, if a hazardous constituent is causing a demonstrated risk to human health or the environment, that constituent must be monitored during the corrective action. Ground water at designated monitor wells is sampled for all hazardous constituents at the end of each major phase of corrective action and again before license termination and transfer of the site to the custodial agency for long-term custody.

For compliance monitoring, after a corrective action program has been terminated, compliance monitoring at the point of compliance will resume for the duration of the compliance period, until license termination, as defined in 10 CFR Part 40, Appendix A, Criterion 7A.

(9)   Design of Surface Impoundments.

The reviewer should verify that any impoundment built as part of the corrective action program to contain wastes is acceptably designed, constructed, and installed. The design, installation, and operation of these surface impoundments must meet relevant guidance in Regulatory Guide 3.11, Section 1 (NRC 1977). Materials used to construct the liner should be reviewed to determine that they have acceptable chemical properties and sufficient strength for the design application. The reviewer should confirm that the liner will not be overtopped. The reviewer should also confirm that a proper quality control program is in place.

The review should ensure that the applicable requirements of 10 CFR Part 40, Appendix A, Criterion 5(A) have been met. If the waste water retention impoundments are located below grade, the reviewer should determine that the surface impoundments have an acceptable liner to ensure protection of ground water. The location of a surface impoundment below grade will eliminate the likelihood of embankment failure that could result in release of waste water. The reviewer should determine that the design of associated dikes is such that they will not experience massive failure.

The design of a clay or synthetic liner and its component parts should be presented. At a minimum, design details, drawings, and pertinent analyses should be provided. Expected construction methods, testing criteria, and quality assurance programs should be presented. Planned modes of operation, inspection, and maintenance should be discussed in the application. Deviations from these plans should be submitted to the staff for approval before implementation.

The liner for a surface impoundment used to manage uranium and thorium byproduct material must be designed, constructed, and installed to prevent any migration of wastes out of the impoundment to the subsurface soil, ground water, or surface water at any time during the active life of the surface impoundment. The liner may be constructed of materials that allow wastes to migrate into the liner provided that the impoundment decommissioning includes removal or decontamination of all waste residues, contaminated containment system components, contaminated subsoils, and structures and equipment contaminated with waste and leachate.

4-51

Protecting Water Resources

The liner must be constructed of materials that have appropriate chemical properties and sufficient strength and thickness to prevent failure caused by pressure gradients, physical contact with the waste or leachate, climatic conditions, and the stresses of installation and daily operation. The subgrade must be sufficient to prevent failure of the liner caused by settlement, compression, or uplift. Liners must be installed to cover all surrounding earth that is likely to be in contact with the wastes or leachate.

Tests should show conclusively that the liner will not deteriorate when subjected to the waste products and expected environmental and temperature conditions at the site. Applicant test data and all available manufacturers test data should be submitted with the application for this purpose. For clay liners, tests, at a minimum, should consist of falling head permeameter tests performed on columns of liner material obtained during and after liner installation. The expected reaction of the impoundment liner to any combination of solutions or environmental conditions should be known before the liner is exposed to them. Field seams of synthetic liners should be tested along the entire length of the seam. Representative sampling may be used for factory seams. The testing should use state-of-the-art test methods recommended by the liner manufacturer. Compatibility tests that document the compatibility of the field seam material with the waste products and expected environmental conditions should be submitted for staff review and approval. If it is necessary to repair the liner, representatives of the liner manufacturer should be called on to supervise the repairs.

Proper preparation of the subgrade and slopes of an impoundment is very important to the success of the surface impoundment. The strength of the liner is heavily dependent on the stability of the slopes of the subgrade. The subgrade should be treated with a soil sterilant. The subgrade surface for a synthetic liner should be graded to a surface tolerance of less than 2.54 cm [1 in.] across a 30.3-cm [1-ft] straightedge. NRC Regulatory Guide 3.11, Section 2 (NRC, 1977) outlines acceptable methods for slope stability and settlement analyses, and should be used for design. If a surface impoundment with a synthetic liner is located in an area in which the water table could rise above the bottom of the liner, underdrains may be required. The impoundment will be inspected in accordance with Regulatory Guide 3.11.1 (NRC, 1980).

To prevent damage to liners, some form of protection should be provided, such as (a) soil covers, (b) venting systems, (c) diversion ditches, (d) side slope protection, or (e) game-proof fences. A program for maintenance of the liner features should be developed, and repair techniques should be planned in advance.

The surface impoundment must have sufficient capacity and must be designed, constructed, maintained, and operated to prevent overtopping resulting from (a) normal or abnormal operations, overfilling, wind and wave actions, rainfall, or run-on; (b) malfunctions of level controllers, alarms, and other equipment; and (c) human error. If dikes are used to form the surface impoundment, they must be designed, constructed, and maintained with sufficient structural integrity to prevent their massive failure. In ensuring structural integrity, the applicant must not assume that the liner system will function without leakage during the active life of the impoundment.

Controls should be established over access to the impoundment, including access during routine maintenance. A procedure should be developed that ensures unnecessary traffic is not directed to the impoundment area. A program should be established to ensure that daily inspections of tailings or waste impoundment systems are conducted and recorded and that failures or unusual conditions are reported to the NRC.

In addition, the reviewer should evaluate the proposed surface impoundment to determine if it meets the definition of a dam as given in Regulatory Guide 3.11 (NRC, 1977). If this is the case, the surface impoundment should be included in the NRC dam safety program, and be subject to Section 215, "National Dam Safety Program," of the Water Resources Development Act of 1996. If the reviewer finds that the impoundment conforms to the definition of a dam, the dam ranking (low or high hazard) should be evaluated. If the dam is considered a high hazard, an emergency action plan is needed consistent with Federal Emergency Management Agency requirements. For low hazard dams, no emergency action plan is required. For either ranking of dam, the reviewer should also verify that the licensee has an acceptable inspection program in place to ensure that the dikes are routinely checked, and that performance is properly maintained.

A quality control program should be established for the following factors: (a) clearing, grubbing, and stripping; (b) excavation and backfill; (c) rolling; (d) compaction and moisture control; (e) finishing; (f) subgrade sterilization; and (g) liner subdrainage and gas venting.

(10) Financial Surety Is Provided.

The licensee must maintain a financial surety, within the specific license, for the cleanup of ground water, with the surety sufficient to recover the anticipated cost and time frame for achieving compliance, before the land is transferred to the long-term custodian. The financial surety must be sufficient to cover the cost of corrective action measures that will have to be implemented if required to restore ground-water quality to the established site-specific standards (including an alternate concentration limit standard) before the site is transferred to the government for long-term custody. Guidance on establishing financial surety is presented in NRC (1988, 1997). Appendix C to this standard review plan provides an outline of the cost elements appropriate for establishing surety amounts for conventional uranium mills. The financial surety review is acceptable if the applicant's assessment and any staff assessment of the surety amounts are reasonably consistent.

4.4.4 Evaluation Findings

If the staff review, as described in standard review plan Section 4.4, results in the acceptance of the ground-water corrective action plan and compliance-monitoring plans, the following conclusions may be presented in the technical evaluation report:

The staff has completed its review of the ground-water corrective action and compliance monitoring plans at the _____ uranium mill facility. This review included an evaluation using the review procedures in Section 4.4.2 and the acceptance criteria outlined in Section 4.4.3 of this standard review plan. The ground-water corrective action program should achieve the goal of returning hazardous constituent concentration levels in ground water to the concentration limits set as standards in 10 CFR Part 40, Appendix A, Criterion 5D. The monitoring program will provide reasonable assurance that, after the corrective actions have been taken, the ground-water protection standard will not be exceeded.

The licensee has established a ground-water compliance strategy that is acceptable for the site. The strategy consists either of no remediation or active remediation when contaminants are present at concentrations above background levels, maximum concentration limits, or alternate concentration limits. When active remediation is necessary, the remedial action design and implementation are acceptable. The licensee has acceptably presented pumping/injection rates, treatment methods, equipment and maintenance requirements, and plans and schedules for construction, and has produced maps showing locations of remediation equipment. An analysis has been conducted that demonstrates (1) the chosen active remediation system technology is appropriate for the site conditions, (2) design pumping rates are sustainable and will control migration of contaminants away from the site, and (3) the natural heterogeneity of the system has been acceptably accounted for in a conservative remediation strategy. The licensee has identified acceptable waste management practices. Qualified engineers, state authorities, and national agencies have provided appropriate oversight. Institutional controls are appropriate for the site, including (1) controlling access to the site, (2) conducting periodic inspections, and (3) periodically monitoring cleanup performance. The monitoring program includes (1) a description of quality assurance procedures; (2) the number of monitoring wells and their locations; (3) a list of constituents that will be sampled, along with the sampling frequency for each monitored constituent; and (4) action levels for triggering enhanced monitoring or revisions to cleanup activities. The licensee has described an acceptable scheme for cleanup and compliance monitoring. The licensee will sample ground water at the point of compliance for all hazardous constituents of concern. An adequate surety mechanism and fund has been established to support the ground-water cleanup.

On the basis of the information presented in the application and the detailed review conducted of the ground-water corrective action and compliance monitoring plans for the _____ uranium mill facility, the NRC staff concludes that the information is acceptable and is in compliance with the following criteria in 10 CFR Part 40, Appendix A: Criteria 5A(4) and 5A(5), which require proper operation of impoundments and design of dikes; Criterion 5B, which requires NRC to establish a list of hazardous constituents, concentration limits, a point of compliance, and a compliance period; Criterion 5C, which provides a table of secondary concentration limits for certain constituents when they are present in ground water above background concentrations; Criterion 5(D), which provides requirements for a ground-water corrective action program; Criterion 5E, which requires licensees conducting ground-water protection programs to consider the use of bottom liners, recycle of solutions and conservation of water, dewatering of tailings, and neutralization to immobilize hazardous constituents; Criterion 5F, which requires that, where ground-water impacts from seepage are occurring at an existing site, action must be taken to alleviate the conditions that lead to seepage, and ground-water quality must be restored, including providing technical specifications for the seepage control system and implementation of a quality assurance program; Criterion 5G, which requires licensees to perform site characterization in support of a tailings disposal system proposal; Criterion 5H, which requires steps be taken during stockpiling of ore to minimize penetration of radionuclides into underlying soils; Criterion 7A, which provides for establishment of three types of monitoring systems: detection, compliance, and corrective action; Criterion 8A, which requires proper inspection and documentation of the operation of tailings and waste retention systems; and Criterion 13, which provides a list of hazardous constituents that must be considered when establishing the list of hazardous constituents in ground water at any site.

If surface impoundments are to be used at the facility to manage byproduct material, the design of dikes used to construct surface-water impoundments has been demonstrated to comply with Regulatory Guide 3.11, Sections 2 and 3 (NRC, 1977) and, therefore, comply with requirements of 10 CFR Part 40, Appendix A, Criterion 5(A)5. In addition, because the impoundment dikes may conform to the definition of a dam as given in the Federal Guidelines for Dam Safety, they are subject to the NRC dam safety program, and to Section 215, "National Dam Safety Program, of the Water Resources Development Act of 1966."

Surety funds and funding methods proposed by the applicant comply with 10 CFR Part 40, Appendix A, Criteria 9 and 10, which establish financial requirements for conventional uranium mills.

4.4.5 References

American Society for Testing and Materials Standards

D 5447, "Standard Guide for Application of a Ground-Water Flow Model to a Site-Specific Problem."

D 5490, "Standard Guide for Comparing Ground-Water Flow Model Simulations to Site-Specific Information."

D 5609, "Standard Guide for Defining Boundary Conditions in Ground-Water Flow Modeling."

D 5611, "Standard Guide for Conducting a Sensitivity Analysis for a Ground-Water Flow Model Application."

D 5718, "Standard Guide for Documenting a Ground-Water Flow Model Application."

D 5880, "Standard Guide for Subsurface Flow and Transport Modeling."

D 5981, "Standard Guide for Calibrating a Ground-Water Flow Model Application."

Hostetler, C.J. and R. L. Erickson. "Coupling of Speciation and Transport Models." *Metals in Groundwater.* H.E. Allen, E.M. Perdue, and D.S. Brown, eds. Chelsea, Michigan: Lewis Publishers. pp. 173–208. 1993.

Murphy, W.M. and E.L. Shock. "Environmental Aqueous Geochemistry of Actinides." *Uranium: Mineralogy, Geochemistry, and the Environment.* P.C. Burns and R. Finch, eds. San Antonio, Texas: Center for Nuclear Waste Regulatory Analyses. 1999.

Protecting Water Resources

Parkhurst, D.L. and A.A.J. Appello. "User's Guide to PHREEQC (Version 2)—A Computer Program for Speciation, Batch-Reaction, One Dimensional Transport, and Inverse Geochemical Modeling." 99-4259. Washington, DC: U.S. Geological Survey. 1999.

Talisman Partners, Ltd. "Introduction to RACER 2000™ (Version 2.1.0). A Quick Reference." Englewood, Colorado: Talisman Partners, Ltd. 2000.

NRC. "Technical Position on Financial Assurances for Restoration, Decommissioning, and Long-Term Surveillance and Control of Uranium Recovery Facilities." Washington DC: NRC. 1988.

————. Regulatory Guide 3.11.1, "Operational Inspection and Surveillance of Embankment Retention Systems for Uranium Mill Tailings." Rev. 1. Washington, DC: NRC, Office of Standards Development. 1980.

————. Regulatory Guide 3.11, "Design, Construction, and Inspection of Embankment Retention Systems for Uranium Mills." Washington, DC: NRC, Office of Standards Development. 1977.

5.0 RADIATION PROTECTION

This chapter of the standard review plan establishes the guidelines for NRC staff to perform and document its review of the proposed radiation protection design for disposal cell covers, for the cleanup of soil and structures contaminated with byproduct material (soil removal, building demolition and disposal or decontamination), and for the proposed radiation safety controls and monitoring during reclamation and decommissioning activities. The radiation standards to be addressed in the evaluation of the reclamation plan include 10 CFR Part 40, Appendix A, Criterion 6(1), which establishes a long-term radon flux limit and direct gamma exposure (background) level for the tailings disposal cell cover, and Criterion 6(5), which requires that the radioactivity of near-surface cover materials be essentially the same as surrounding surface soils. Also, the decommissioning plan, whether submitted as part of the reclamation plan or provided in detail as a separate document, should comply with 10 CFR 40.42(g)(4) and (5,) which requires a description or procedures indicating how the licensee will demonstrate that the residual radioactivity levels in land and on structure surfaces meet Appendix A, Criterion 6(6) (see Appendix H guidance in this standard review plan on the radium benchmark dose approach for cleanup of residual radionuclides other than radium). In the review, the staff should consider any licensee-proposed alternatives to Appendix A criteria as described in the Introduction of Appendix A to 10 CFR Part 40.

5.1 Disposal Cell Cover Radon and Gamma Attenuation and Radioactivity Content

5.1.1 Areas of Review

The areas of review for radon attenuation (radon barrier design) are the radiological and physical properties of the contaminated and cover materials and the application of the computer code or other methods used for calculating the estimated long-term radon flux from the completed disposal cell. The areas of review for the control of gamma radiation from the disposed waste and for the radioactivity content of the cover are the proposed methods to demonstrate compliance with the regulations. This area would also include consideration of disposal of wastes from processing alternate feed materials and non-11e.(2) byproduct material in uranium mill tailings impoundments, if such action is proposed.

The radon barrier portion of the disposal cell cover is the layer or layers designed to reduce radon flux from the cell. Other cover layers contribute to radon attenuation and may be considered in the flux calculation.

For the radon barrier design, the staff should review (1) the bases, assumptions, and procedures for determining the input parameter values of the tailings and other wastes and radon barrier materials (such as the sampling and testing programs); (2) procedures for materials placement in the disposal cell, as presented in the reclamation plan construction specifications; (3) the description of the model (numerical or analytical) used to approximate the average long-term radon flux at the cover surface; and (4) if the standard computer codes for estimating radon flux (RADON, RAECOM) are not used, references for the methodology used to calculate the long-term radon flux from the cover.

Radiation Protection

For cover gamma attenuation, the staff reviews the proposed procedure to calculate or measure the gamma level (exposure rate or count rate) on the cover. For the radioactivity content, the staff should review the proposal for measurements in the upper 61 cm [2 ft] of cover. Alternatively, the staff should review proposed control measures on the cover material before placement to demonstrate that the average radioactivity content of this layer is not distinguishable from local surface soil and to demonstrate that it does not include waste or rock containing elevated levels of radium.

5.1.2 Review Procedures

5.1.2.1 Radon Attenuation

The radon barrier design, as presented in the reclamation plan, should be reviewed along with the data supporting the design. Chapter 2.0 of this standard review plan presents review areas, procedures, and acceptance criteria for geotechnical information related to material properties and cell stability. The staff members assigned the health physics and geotechnical reviews should coordinate the review of the radon attenuation design and analysis. The geotechnical properties of the cover layers will be considered in the context of their influence on the cover integrity by considering long-term moisture content of the radon barrier. Materials underlying the radon barrier are evaluated for stability so that the cover will not experience cracking from settlement or subsidence, as discussed in Chapter 2.0.

In addition, the health physics reviewer should:

(1)   Evaluate the basis for selection of parameter values for tailings and cover material properties to determine if the values are based on appropriate measurements or estimates and will lead to a reasonably conservative estimate of the radon flux. The scope and techniques used for site investigations should be examined to ensure that the field investigation (boring, sampling, and surveying) and testing programs will produce representative data needed to support the conclusions of the analyses.

(2)   Assess whether parameter values are consistent with anticipated construction specifications and reflect expected long-term conditions at the site. The radon flux estimate must represent the average flux, for periods of more than 1 year but less than 100 years, and consider that the cell design life is 1,000 years.

(3)   Determine whether the parameter values reflect the meteorological and hydrological conditions at the disposal site, bulk density, type of material, and the influence of overlying material layers. The cover material moisture content must be determined by accurately measured values or reasonably conservative estimates. Preferably more than one method is utilized, as there are limitations to each method and the long-term moisture content of the radon barrier is one of the most important parameters in the flux model.

(4)   Determine that the radium (Ra-226) activity concentration in picocuries per gram (pCi/g) within the tailings cell has been, or will be, measured directly from representative tailings

samples, and other large-volume sources of contaminated material, utilizing an acceptable method. If the tailings were placed so that specific areas in the pile contain higher Ra-226 content (e.g., slime tailings), then Ra-226 values and the modeling should represent the layering or localization of the significantly elevated Ra-226 levels in the upper 3.6 m [12 ft], as deeper material generally has little effect on the radon flux. This approach is necessary because modeling higher concentrations of Ra-226 in the upper few feet of the pile would result in a higher radon flux estimate than using an average Ra-226 value for the entire upper 3.6 m [12 ft]. Also, if large quantities of material containing thorium-230 (Th-230) levels significantly higher than the Ra-226 levels are placed in the upper portion of the pile, the 1,000-year Ra-226 concentration (Ra-226 remaining from the residual Ra-226 and from the decay of Th-230) should be considered for that layer of material in the modeling.

In accordance with Footnote 2 of Criterion 6(1), the radon emissions from covering materials should be estimated as part of developing a closure (reclamation and decommissioning) plan. If any layer of the cover will contain material with above-background levels of Ra-226 or Th-230, the licensee should model that layer with a conservatively high estimated Ra-226 level, or should commit to measure the cover radionuclide level(s) during or after placement to confirm the adequacy of the radon attenuation design. A commitment from the licensee to confirm the cover Ra-226 content in the reclamation completion report should be present if the borrow site measurements are limited and the possible cover Ra-226 level could prevent the radon flux from being in compliance.

(5)   Evaluate each code input parameter value, keeping in mind that the code default parameter values are not always conservative, and then consider the set of parameter values as a whole (balance of conservatism and uncertainty). It is the total flux model that will be approved, not individual parameter values. Consider that the void ratio, the density, porosity, and moisture saturation values should be typical of the soil type in each layer of the cell. The radon flux model should result in a representative and a reasonably conservative (given the uncertainty in some values) long-term radon flux estimate.

(6)   A measured, not a calculated, disposal cell average radon flux is required by Appendix A, Criterion 6(2), as soon as practical after placement of the radon barrier, and Criterion 6(3) stipulates that radon-222 release rates must be verified for each portion of the pile or impoundment as the final radon barrier for that portion is placed, when phased emplacement of the final radon barrier is included in the applicable reclamation plan required by 10 CFR Part 40, Appendix A, Criterion 6A(1). Therefore, the reviewer should document in the technical evaluation report whether the reclamation plan stipulates if the radon barrier is to be placed in phases or as a fairly continuous operation. In either case, the final radon barrier must be placed as expeditiously as practicable. However, some tailings cells have evaporation ponds on top that can't be covered until the ground-water correct action is complete. A commitment to measure and document the radon flux on the final radon barrier, as required by Criterion 6(2) and (4), should be in the reclamation plan. Before the measurements are performed, a map of the disposal cell indicating the measurement locations and outline of tailings and

cover extent should be reviewed by the NRC staff before the measurements are performed.

(7)   Guidance on the disposal of wastes from processing alternate feed materials and non-11e.(2) byproduct materials in uranium mill tailings impoundments is presented in Appendix I to this standard review plan. The staff should use this guidance when evaluating requests to dispose of such materials and consider their impact on the radon attenuation design.

5.1.2.2   Gamma Attenuation

Most radon barriers should be thick enough to reduce the gamma level of the disposal cell to background. To demonstrate compliance with this aspect of Criterion 6(1), the cover gamma attenuation can be calculated based on the shielding value of the cover soil. Alternatively, the licensee commits to (1) measure the gamma level at 1 meter above the completed cover (or radon barrier) with at least one measurement per acre and (2) demonstrate that the average gamma level for the cell is comparable to the local background value.

5.1.2.3   Cover Radioactivity Content

At some mill facilities, uranium deposits, open pit uranium mines, overburden piles (soil moved from the pit area), and/or reclaimed mining areas are on or near the site. All of these areas would contain elevated levels of uranium, radium, and the other radionuclides in the uranium decay chain. In determining what surrounding soil values may be compared to the radionuclide content of the disposal cell cover, the mining areas reclaimed/restored under state regulations may be included. Also, consideration of the low health risk of human exposure to the cell cover and the perpetual custody of the cell by the government may be part of the risk-informed approach. If the average radioactivity (mainly radium level) for the cover material exceeds the average value for surrounding soil, the reclamation plan should contain a statistical analysis of the distributions of surrounding soil (not necessarily undisturbed background) and cover radioactivity to demonstrate that they are not significantly different.

5.1.3   Acceptance Criteria

5.1.3.1   Radon Attenuation

The radon attenuation design will be acceptable if it meets the following criteria:

(1)   The one-dimensional, steady-state gas diffusion theory for calculating radon flux and/or minimum cover thickness is used. An acceptable analytical method for determining the necessary cover thickness to reduce radon flux to acceptable limits or to determine the long-term radon flux from the proposed cover is the computer code RAECOM (NRC, 1984) and the comparable RADON code (NRC, 1989). The main difference between the two codes is that RADON does not have the optimization for cost benefit calculations. The staff will use the RADON code to verify the analysis. Other methods that estimate the average surface radon release from the covered tailings may be

acceptable, if it can be shown that these methods produce reliable estimates of radon flux.

(2)   With the RAECOM and RADON computer codes, the radon concentration above the top layer is either set to a conservative value of zero or a measured background value is used. The precision number (the level of computational error that is acceptable) is set at 0.001.

(3)   The estimates of the material parameters used in the radon flux calculations are reasonably conservative, considering the uncertainty of the values. For all site-specific parameters, supporting information describing the test method and its precision, accuracy, and applicability is provided. The basis for the parameter values and the methods in which the values are used in the analyses are adequately presented. Moisture-dependent parameter values (e.g., radon emanation coefficient and diffusion coefficient) are based on the estimated long-term moisture content of the materials at the disposal site.

The materials testing programs employ appropriate analytical methods and sufficient and representative samples were tested to adequately determine material property values for both cover soils and contaminated materials. In the absence of sufficient test data, conservative estimates are chosen and justified. The quality assurance program for parameter data is adequate and the data are available for inspection. All parameter values are consistent with anticipated construction specifications and represent expected long-term conditions at the site.

(4)   The contaminated material thickness is determined from estimates of total tailings production or waste placement and the areal extent, from boring logs, or changes in elevation from pre- to post-operation. Either the estimated thickness of a tailings source is used, or alternatively, the RADON code default value of 500 cm [16.4 ft] is used (NRC, 1989).

(5)   Dry bulk densities of the cover soils and tailings material are determined from Standard Proctor Test data (American Society for Testing and Materials D 698) or Modified Proctor Test data (American Society for Testing and Materials D 1557). Radon barrier materials are usually compacted to a minimum of 95 percent of the maximum dry density as determined by American Society for Testing and Materials D 698 or to a minimum of 90 percent of the maximum dry density as determined by American Society for Testing and Materials D 1557. Field or placement densities to be achieved based on the construction specifications are used in the calculations. If the pile is stabilized in place, the *in situ* bulk density for the tailings is used in the analysis.

Porosities are measured by mercury porosimetry or another reliable method, or the method for estimating the porosity of cover soils and tailings materials using the bulk density and specific gravity given in Regulatory Guide 3.64 (NRC, 1989) is used.

If a portion of the modeled cover (radon attenuation layers) could be affected by freeze-thaw events, that portion is represented in the model with lower density and corresponding higher porosity values than the unaffected portion. The U.S. Army Corps of Engineers (1988) and the DOE (1988) have demonstrated that freeze-thaw cycles can increase the permeability of compacted clay by 40 to 300 times the original value. For fine-grained soils with some sand (50-percent fines), the DOE conservatively estimated that freeze-thaw cycles could lower the density by 14 percent (DOE, 1992). Also see the discussion in Section 2.5.3 of this standard review plan.

(6)   The long-term moisture content that approximates the lower moisture retention capacities of the materials or another justified value is used. Estimated values for the long-term moisture content can be compared with present *in situ* values to assure that the assumed long-term value does not exceed the present field value. Borrow samples can be taken at a depth of 120 to 500 cm (3.9 to 16.4 ft), but not close to the water table, and the borrow site conditions should be correlated to conditions at the disposal site.

The following methods are acceptable for estimating the long-term soil moisture, but each has limitations:

(a)   Laboratory procedures American Society for Testing and Materials D 3152 (fine-textured soils) and American Society for Testing and Materials D 2325 (coarse and medium-textured soils) for capillary moisture test (15-bar suction) corresponding to the moisture content at which permanent wilting of plants occurs (Baver, 1956).

(b)   The empirical relationship (Rawls and Brakensiek, 1982) that predicts water retention values of a soil on a volume basis (appears to be more suitable to sandy and silty soil than to clayey soil) and is represented by:

$$c = 0.026 + 0.005x + 0.0158y$$

where

c = predicted 15-bar soil water-retention value (volumetric moisture content)
x = percent clay in the soil
y = percent organic matter in the soil

This method takes into consideration the particle-size distribution of the soil. Clay particle sizes are defined here as those less than 0.002 mm in diameter. Organic content measurement is generally determined by reaction with hydrogen peroxide or by exposure to elevated temperature. The volumetric moisture content value derived from this equation should be converted to a weight percentage for application in the RAECOM and RADON codes. Other empirical correlations (Section 7.1.3 of DOE, 1989), if adequately justified, may be acceptable.

(7)   Values for Ra-226 activity (pCi/g) are measured directly from tailings samples and other large volume sources of contaminated material, by radon equilibrium gamma spectroscopy (allow at least 10 days for the sealed sample to equilibrate), wet chemistry

alpha spectrometry, or an equivalent procedure. If the tailings are fairly uniform in Ra-226 content and the Ra-226 and uranium (U-238) in the ore were approximately in equilibrium, the Ra-226 activity can be estimated from the average ore grade processed at the site, as discussed in Regulatory Guide 3.64 (NRC, 1989). Generally, tailings should be sampled at 90-cm [3-ft] intervals to a depth of 366 cm [12 ft], including representative sampling of slime tailings. More than one layer of contaminated material is represented in the flux model if there are significant differences in Ra-226 content with depth.

Since the disposal cell performance standard deals only with radon generated by the contaminated material, it is acceptable to neglect the Ra-226 activity in the cover soils for modeling flux, provided the cover soils are obtained from materials not associated with ore formations or other radium-enriched materials. If deep {below 61 cm [2 ft]} cover layers contain elevated Ra-226 or Th-230, that material layer and its Ra-226 level is represented in the flux model.

(8)   The emanation coefficient has been obtained by using methods provided in Nielson, et al. (1982) and properly documented, or otherwise set to the reasonably conservative (for most soils) code default value of 0.35. A value of 0.20 may be estimated for tailings based on the literature, if supported by limited site-specific measurements.

(9)   The radon diffusion coefficient, D, represents the long-term properties of the materials. The D value is determined from direct measurements or appropriately calculated. The soil should be tested at the design compaction density, with a range of moisture content values that includes the lower moisture retention capacity of the soil so that a radon breakthrough curve can be obtained (DOE, 1989). The calculation of the diffusion coefficient, based on the long-term moisture saturation, and porosity, as proposed in Regulatory Guide 3.64, Section C.1.1.5 (NRC, 1989), and the optional calculation in the RADON code, are acceptable.

(10)   The soil cover thickness proposed in the reclamation design is such that the calculated average long-term radon flux is reduced to a level that meets the requirement in 10 CFR Part 40, Appendix A, Criterion 6(1).

5.1.3.2   Gamma Attenuation

The proposed cover will reduce the gamma radiation from the byproduct material to local soil background levels, and the licensee proposed an acceptable method to demonstrate this. The data will appear in the reclamation completion report.

5.1.3.3   Cover Radioactivity Content

At least the upper 61 cm [2 ft] of the disposal cell cover will contain levels of radioactivity essentially the same as surrounding soils, as demonstrated by an appropriate procedure. The data will be in the reclamation completion report if not available for the reclamation plan.

5.1.4 Evaluation Findings

If the staff review, as described in this section, results in the acceptance of the radon and gamma attenuation and cover radioactivity content assessments, the following conclusions may be presented in the technical evaluation report:

The staff has completed its review of the disposal cell cover radiation control at the _____ uranium mill facility. This review included an evaluation using the review procedures in Section 5.1.2, and the acceptance criteria outlined in Section 5.1.3 of this standard review plan.

The licensee has presented an acceptable radon attenuation design, and the staff evaluation determines that (1) the method used for calculating radon flux or minimum cover thickness is based on the one-dimensional, steady-state gas diffusion theory and appropriate input values; (2) input values of the material parameters lead to a reasonably conservative estimate of the long-term radon flux; (3) material parameters are consistent with construction specifications and expected long-term conditions; (4) the long-term attenuating capability of cover materials is justified using acceptable results of relevant tests or conservative estimates; (5) estimates of contaminated materials thickness are determined utilizing a sufficient number of data or by use of the default value; (6) if not measured, the estimated porosity of cover soils and tailings materials is based on the method in Regulatory Guide 3.64; (7) soil moisture values represent long-term moisture retention capacities; (8) Ra-226 activity has been measured in the tailings and other large volume sources of contaminated materials using acceptable procedures; (9) the emanation coefficient is obtained by either the equilibration method or the prediction method, or is set to a reasonably conservative value of 0.35; (10) the radon diffusion coefficient of the cover soil is determined from direct measurements or from a calculation based on Regulatory Guide 3.64; and (11) the cover gamma level and radioactivity content will be correctly determined and documented.

On the basis of the information presented in the application and in detailed review conducted of the site characterization for the _____ uranium mill facility, the NRC staff concludes that the disposal cell cover radon and gamma attenuation and radioactivity content are in compliance with 10 CFR Part 40, Appendix A, Criterion 6(1), which requires placement of an earthen cover (or approved alternative) over tailings and wastes at the end of the milling operations while providing assurance of control of radiological hazards for 1,000 years, to the extent reasonably achievable (but no less than 200 years); and which limits releases of radon-222 from uranium byproduct materials to the atmosphere so as not to exceed an average rate of 20 picocuries per square meter per second (pCi/m^2-s); Criterion 6(2), which requires demonstration of the effectiveness of the final radon barrier prior to emplacement of erosion protection measures or other features; Criterion 6(3), which requires demonstration of the effectiveness of phased placement of radon barriers as each phase is completed if phased placement is in the plan; and Criterion 6(5), which requires that radon exhalation is not significantly above background because of the cover material.

5.1.5 References

American Society for Testing and Materials Standards:

D 698-91, "Test Method for Laboratory Compaction Characteristics of Soil Using Standard Effort."

D 1557-91, "Test Method for Laboratory Compaction Characteristics of Soil Using Modified Effort."

D 2325-68, "Standard Test Method for Capillary-Moisture Relationships for Coarse- and Medium-Textured Soils by Porous-Plate Apparatus."

D 3152-72, "Standard Test Method for Capillary-Moisture Relationships for Fine-Textured Soils by Pressure-membrane Apparatus."

Baver, L.D. *Soil Physics*. New York, New York: John Wiley and Sons. pp. 283–303. 1956.

DOE. 1988. "Effect of Freezing and Thawing on UMTRA Covers." Albuquerque, New Mexico: DOE, Uranium Mill Tailings Remedial Action Project. 1988.

———. "Technical Approach Document." UMTRA–DOE/AL–050425.0002. Revision II. Albuquerque, New Mexico: DOE. December 1989.

———. "Remedial Action Plan and Site Design for Stabilization of the Inactive Uranium Mill Tailings Site at Gunnison, Colorado." Remedial Action Selection Report. UMTRA–DOE/AL–050508. Albuquerque, New Mexico: DOE. October 1992.

Nielson, K.K., et al. "Radon Emanation Characteristics of Uranium Mill Tailings." Proceedings of the Symposium on Uranium Mill Tailings Management December 9–10. Ft. Collins, Colorado: Colorado State University. 1982.

NRC. NUREG/CR–3533, "Radon Attenuation Handbook for Uranium Mill Tailings Cover Design." Washington, DC: NRC. 1984.

———. Regulatory Guide 3.64, "Calculation of Radon Flux Attenuation by Earthen Uranium Mill Tailings Covers." Washington, DC: NRC. 1989.

Rawls, W.J. and D.L. Brakensiek. "Estimating Soil Water Retention From Soil Properties." Proceedings of the American Society of Civil Engineers. *Journal of the Irrigation and Drainage Division*. Vol. 108, No. IR2. 1982.

U.S. Army Corps of Engineers. "Effects of Freezing and Thawing on the Permeability of Compacted Clay." Hanover, New Hampshire: Cold Regions Research and Engineering Laboratory. 1988.

Radiation Protection

5.2   Decommissioning Plan for Land and Structures

5.2.1   Areas of Review

The areas of review for the decommissioning (radiological cleanup and restoration) of land and structures (e.g., towers and buildings) are the site conditions (nature and extent of the contamination, soil background radioactivity, etc.); planned decommissioning activities (how and what measurements will be made, quality assurance quality control program, gamma guideline levels for soil cleanup, how "as low as is reasonably achievable" will be demonstrated); methods to be used to protect workers, the public, and the environment; verification (final status survey) plan with procedures; and the decommissioning cost estimate and surety amount. Often, the detailed mill decommissioning plan and the soil cleanup and verification plan are submitted for NRC approval a year before decommissioning is scheduled to begin. However, the reclamation plan must describe the expected decommissioning activities in enough detail to support the cost estimate needed for surety purposes. The preliminary decommissioning plan in the reclamation plan should include commitments to provide a detailed plan and cost estimate for NRC approval at least 9 months before decommissioning is expected to begin.

5.2.2   Review Procedures

(1)   Site Conditions (Characterization)

Based on the operational history (including radiation surveys) of the facility, the reviewer determines that the plan describes the likely source and locations of residual byproduct material such as spills, releases, waste burial, haul roads, diversion ditches, process and yellowcake storage areas, ore stockpile areas, areas likely to be affected by windblown tailings, and tailings solution evaporation ponds. Determine that the extent of contamination (area and depth for soil) has been or will be established from adequate representative sampling and surveying. Sample analysis should include uranium where yellowcake or ore dust was present and thorium (Th-230) for acidic tailings pond residue. The radiological analysis for the ore processed at the site should also be reviewed for the ratios of Ra-226/U-238 and Ra-226/Th-230 to determine if non-equilibrium conditions could exist in the contaminated soil. The U-238 activity can be estimated by dividing the U-nat (total uranium) value by two. The reviewer should also determine from this data if Th-232 could be elevated above background due to windblown tailings and whether additional characterization data should be provided.

(2)   Soil Background Radioactivity

Determine that the background level of Ra-226 (and U-nat, Th-230 and Th-232, as needed) in surface {15 cm [6 in.]} soil has been estimated using representative soil samples from nearby {within 3.2 km [2 mi] of site boundary} undisturbed areas that are not affected by site activities and are geologically and chemically similar to the contaminated areas. The number of samples will depend partly on the variability in background values, but at least 30 samples should be obtained at the typical site to determine the average value, standard deviation, and distribution. The arithmetic mean of the sample data is used in the cleanup criteria unless appropriate statistical analysis demonstrates a log normal distribution (three tests) of the data.

Several different background values may be required if contaminated areas have distinctly different soil types. For example, if a portion of the site has a natural uranium and/or radium mineralization zone in/near the surface, the cleanup criterion for that area would use a background (reference) U-238 or Ra-226 value from a similarly mineralized area. A geologic site map with the background values placed on the sample location can be used to help identify whether more than one background value should be considered.

If the plan indicates that *in situ* ore is in the clean-up area, it should be characterized by Ra-226/U-238 ratios, visual criteria, and/or other means.

(3)   Cleanup Requirements

For land cleanup, the residual Ra-226 [and/or Ra-228 if thorium (Th-232) byproduct material is present] in soil must meet the concentration limits in 10 CFR Part 40, Appendix A, Criterion 6(6), in areas that are not evaluated by the radon flux criterion (i.e., areas other than the disposal cell). If the plan indicates that the subsurface 15 pCi/g Ra-226 standard will be used, its use should be justified. For structures to remain on site, the staff reviews the proposed cleanup of mill-related radionuclides (byproduct material) on surfaces (e.g., walls, floors, drains) as well as in underlying soil.

For NRC uranium recovery licensees that did not have a decommissioning plan approved by June 11, 1999, [Appendix A, Criterion 6(6) was expanded effective that date], or that subsequently submit a revised plan, the radium benchmark dose applies for cleanup of residual radionuclides other than radium [primarily uranium (U-nat) and thorium (Th-230)] in soil and for surface activity on structures. For such licensees, the reviewer should refer to Appendix H of this document for guidance on the benchmark approach. This approach would also be evaluated if proposed by other uranium recovery licensees to derive cleanup limits in order to demonstrate compliance with 10 CFR 40.42(k)(2).

Determine that the plan indicates that residual contamination will be reduced to "as low as is reasonably achievable levels." Usually, a low gamma guideline level is chosen so that most grids are cleaned to near background levels of radiation, an approach that has proven less costly for licensees than more extensive soil sampling and analysis. It is a method acceptable to the staff to demonstrate compliance with the "as low as is reasonably achievable" principle. The administrative limit for surface activity (10 to

Radiation Protection

25 percent of the criteria) has been considered an "as low as is reasonably achievable" level in the past but current policy should be confirmed by staff. The "as low as is reasonably achievable" approach discussed in NUREG–1727 (NRC, 2000b), also may be considered.

(4)   Gamma Guideline Level

Because gamma measurements (in terms of exposure or count rates) can substitute for some Ra-226 analyses [as recommended in 40 CFR 192.20(b)(1)] and such measurements are not very reliable, the reviewer must be sure that the proposed gamma guideline value is conservative, considering the measurement uncertainties involved. Determine that the radium-gamma correlation that is used to derive the gamma guideline was performed with at least 30 soil Ra-226 values from approximately 2 to 25 pCi/g and that the corresponding gamma values adequately represent the grid (100-square-meter area) sampled. The proposed gamma guideline level must reliably (95 percent confidence) result in grids meeting the 5 pCi/g [0.19 Bq/g] Ra-226 plus background standard.

Confirm that the plan contains a commitment to perform a radium-gamma correlation on the verification data, to track soil samples that fail the Ra-226 criteria, and to perform additional cleanup after a verification soil sample exceeds the Ra-226 standard. Just cleaning the failed grid is not adequate because the failed sample could indicate that the gamma value may not be conservative and that some of the unsampled grids may also fail to meet the standard. For example, the plan could indicate that neighboring grids would also be analyzed for Ra-226 or, if the number of failed grids is excessive, the gamma guideline would be adjusted downward and areas further remediated, as necessary.

(5)   Instruments and Procedures

Determine that the instruments and procedures used to determine the soil background radioactivity and the radium-gamma correlation are the same or very similar to those proposed for verification of compliance with Criterion 6(6) (final status survey). See NUREG–1505, Section 4.5 (NRC, 1998a). Instrument sensitivity should be adequate to reliably identify the proposed guideline levels [NUREG–1507 (NRC, 1998b)]. Survey instruments are specified and will be properly calibrated and tested, including daily checks during operations. The reviewer considers national standards (American Society for Testing and Materials, American National Standards Institute, and National Council on Radiation Protection as listed in Section 5.2.5) and the "Multi-Agency Radiation Survey and Site Investigation Manual" [NUREG–1575 (NRC, 2000a)] that contains general principles of soil sampling, determination of background, and gamma surveying to be acceptable.

Soil samples for uranium recovery sites can be composite samples (5 to 11 samples per grid have been approved). Evaluate sampling procedures for completeness (ensure proper depth, identification of sample and location, cleaning of equipment, chain-of-custody, etc). Determine that soil preparation procedures indicate that rocks and vegetation should not be included in the sample to the extent that the additional volume would dilute the soil sample. Generally, rocks greater than or equal to 1.27 cm [0.5 in.] in diameter are excluded. Acceptable sample mixing, drying, and splitting methods are specified.

Evaluate the methods for soil radionuclide analysis. Standard analytical methods should be used. Portions of each sample verifying compliance should be archived until the NRC approves the decommissioning completion (final survey) report, as staff may want to do confirmatory analysis on selected soil samples. The plan for the final disposal of these archived samples should also be reviewed.

As required by 10 CFR 40.42(j)(2)(i), the gamma levels to be reported in the final survey are as mSv (μR) per hour at 1m [39.4 in.] from the surface. Measurements at 1 m [39.4 in.] would allow calculation of an exposure dose, but the goal of the gamma survey is to demonstrate compliance with the radium in soil criterion. Therefore, the staff has approved alternate methods such as meter readings (counts/minute) taken near the ground or at 0.45 m [18 in.]. These methods improve the quality of the gamma-radium correlation by reducing "shine" and they allow the survey meter and equipment associated with a global positioning system to be mounted on an all-terrain vehicle. Typically, measurements are made over the spot to be sampled, or the grid (100 m^2) is scanned with 9 to 12 measurements. Integrated count rate gamma scan values have also been approved if taken for at least one (1) minute within each grid.

Determine that gamma survey procedures indicate the speed, pattern, and spacing of the measurements or scan path. The procedure should allow demonstration of compliance with the radium standard. The reviewer should consider the thoroughness of the gamma scan (remedial action survey) to be done during soil removal, such as 1.5-m [5-ft] scan path, when evaluating the final survey procedures and the percentage of grids proposed for soil sampling.

Determine that procedures for measuring alpha or beta-gamma radiation on structure surfaces are detailed, reflect industry standards, and consider that smears for alpha activity generally have an efficiency of 10 percent or less. Measurements of smears are difficult to interpret quantitatively and should not be used for determining compliance but for determining if further investigation is necessary [NUREG–1575, Sections 6.4.2 and 8.5.3 (NRC, 2000a)].

(6)   Quality Assurance and Quality Control

Determine that the quality assurance/quality control program addresses all aspects of decommissioning. The plan should indicate a confidence interval or that one will be specified before collection of samples. At least 10 percent of the soil samples should be split and a portion sent to an outside laboratory for quality assurance. To properly assess the adequacy of radiological data, the uncertainties associated with the data should be estimated statistically [NUREG–1501, Sections 3.2 and 5.2 (NRC, 1994)].

Evaluate the criteria for validating that the data to be used to demonstrate compliance and the quality assurance procedures to confirm that compliance data are precise and accurate (e.g., laboratory will analyze spiked and duplicate samples, etc.). Confirm that management will ensure that approved procedures are followed (e.g., commitment to check gamma surveyor and data management).

(7)   Final Status Survey

Evaluate the details of the proposed final status survey (radiation surveys and soil analyses) as discussed in Items 3–6 above, and determine whether the survey plan complies with 10 CFR 40.42(j)(2). The reviewer should also determine that enough data of the proper quality can be provided after decommissioning to demonstrate compliance with Criterion 6(6) of Appendix A and 10 CFR 40.42(k)(2). For example, determine that the proposed number and pattern of grids to be soil sampled and analyzed for Ra-226 are justified. Based on the degree of uncertainty (level of error in the measurements, number of measurements), the gamma guideline level, and implementation procedures, the staff has considered soil samples from 0.5 to 10 percent of the grids acceptable. Some verification soil sampling and surveying should be planned in presumably uncontaminated areas (buffer zone of about 30 meters beyond excavated areas). [Refer to Section 3 in Inspection Procedure 87654 (NRC, 2002) for additional information.]

Confirm that the licensee proposes to use the same instruments and procedures for the verification (final status) survey as were used in determining background and for the radium-gamma correlation, or justifies that they are comparable.

If buildings or other structures are to remain on site after license termination, determine whether adequate measurement of the surface activity is planned. Preliminary modeling by staff has indicated, that for habitable buildings, the average total (fixed plus removable) alpha level should be below 2,000 dpm/100 cm^2 in order to achieve 0.25 mSv/yr [25 mrem/yr].

(8)   Preliminary Versus Final Decommissioning Plan

A preliminary decommissioning plan shall be submitted with the reclamation plan and may be updated for license renewal in order for the staff to better evaluate the decommissioning cost estimate provided for surety purposes. Since the actual site decommissioning may be years in the future and continued operation could change the cleanup design, or evolving technology and Agency rules or guidance could change the evaluation of procedures, the review of the preliminary plan should not be technically rigorous. However, the reviewer should determine whether sufficient detail has been provided in the plan to justify that the surety amount for decommissioning activities is adequate.

Confirm that both the preliminary and final plans identify a location to keep the records of information important to the decommissioning, as required by 10 CFR 40.36(f). These records would include documentation of spills or cleanup of contamination, drawings or descriptions of modification of structures in the restricted area, and locations of possible inaccessible contamination.

When a final decommissioning plan is submitted, the reviewer should determine whether the plan addresses the technical aspects discussed above [basically 10 CFR 40.42(g)(4) requirements] and whether it indicates that decommissioning will be completed as soon as practicable. The reviewer follows Section 5.3 of this standard review plan for the evaluation of the health and safety protection aspects of decommissioning. The reviewer should also consider recommendations in Regulatory Guide 3.65 (NRC, 1989) during the evaluation of the final decommissioning plan.

(9)   Non-Radiological Hazardous Constituents

The decommissioning plan must address the non-radiological hazardous constituents of the byproduct material according to 10 CFR Part 40, Appendix A, Criterion 6(7). For windblown tailings areas, meeting the surface Ra-226 standard should be adequate to control these constituents in soil. A tailings cell cover that meets Appendix A criteria should control, minimize, or eliminate postclosure escape of non-radiological hazardous constituents into surface water and the atmosphere. However, any unusual or extenuating circumstances related to such constituents should be discussed in the reclamation plan or decommissioning plan in relation to protection of public health and the environment and should be evaluated by staff. The control of these substances in ground water is addressed under Chapter 4.0 of this standard review plan.

(10)   Decommissioning Cost Estimate

Determine whether the cost estimate is itemized in sufficient detail such that values for soil sampling and preparation, Ra-226 analysis, gamma surveying, and data management are presented. The items should reflect the proposed activities in the plan. Also, the basis for each cost should be provided and verified by staff to be within the range of current charges for such activities in the site region. The staff should verify that adequate surety funds will be provided to cover these costs. Guidance on cost estimates for sureties is provided in Appendix C of this standard review plan.

5.2.3 Acceptance Criteria

The decommissioning plan will be acceptable if it meets the following criteria:

(1)   The plan contains procedures to identify and place within the disposal cell, all soils on and adjacent to the processing site that are in excess of the standards in 10 CFR Part 40, Appendix A, Criterion 6(6), due to site activities. The plan is substantiated by the radiological characterization data and site history.

(2)   Appropriate soil background values (different geological areas may need separate background values) for Ra-226, and for U-nat, Th-230, and/or Th-232, as appropriate, have been proposed with supporting data.

(3)   If elevated levels of uranium or thorium are expected to remain in the soil after the Ra-226 criteria have been met, the licensee has used the radium benchmark dose approach in Appendix H for developing decommissioning criteria for these radionuclides.

Radiation Protection

(4)   To ensure consistency of measurement data, instrumentation and procedures used for soil background analyses and the radium-gamma correlation are the same or very similar to those proposed to provide verification data. The instrumentation has the appropriate sensitivity, and procedures are adequate to provide reliable data.

(5)   A detailed quality assurance and quality control plan for all aspects of decommissioning is provided. In addition to the basis for accepting or rejecting data, a procedure for sampling additional grids when a verification Ra-226 sample fails to meet the standard is provided.

(6)   Final verification (status survey) procedures are adequate to demonstrate compliance with the soil and structure cleanup standards. Survey instruments are specified and will be properly calibrated and tested. The proposed verification soil sampling density takes into consideration detection limits of sample analyses, the extent of expected contamination (unaffected area could have fewer measurements than affected areas), and limits to the gamma survey for the potentially contaminated area to be sampled. The gamma guideline value to be used for verification has been appropriately chosen. Also, there is a commitment to provide the verification soil radium-gamma correlation and the number of verification grids that had additional removal because of excessive Ra-226 values, to confirm that the gamma guideline value was adequate. The plan provides for adequate data collection beyond the excavation boundary (buffer zone).

For structures to remain onsite, adequate plans/procedures to demonstrate compliance with the limits for the surface activity dose in Appendix H of this standard review plan have been developed.

(7)   The plan indicates the location of records important to decommissioning procedures for protection of health and safety and demonstrates that decommissioning will be completed as soon as practicable, as required by 10 CFR 40.42 and Appendix A, Criterion 6A.

(8)   The decommissioning cost estimate is itemized in sufficient detail and a basis (source) for each cost is provided. The total cost is reasonable for the area of the site and the expected decommissioning activities.

(9)   The plan adequately describes the control of non-radiological hazards associated with the wastes as required by 10 CFR Part 40, Appendix A, Criterion 6(7).

(10)   As required by Appendix A, Criteria 9 and 10, the licensee must maintain a financial surety, within the specific license, for the surface reclamation and decommissioning, with the surety sufficient to recover the anticipated cost and time frame for achieving compliance, and include the long-term surveillance. Guidance on establishing financial surety is presented in NRC (1988, 1997). Appendix C to this standard review plan provides an outline of the cost elements appropriate for establishing surety amounts for conventional uranium mills. Any staff assessment of surety amounts is reasonably consistent with the applicant's assessment.

5.2.4 Evaluation Findings

If the staff review, as described in this section, results in the acceptance of the processing site (soil and structures) decommissioning plan, the following conclusions may be presented in the technical evaluation report:

The staff has completed its review of the site decommissioning plan for soil and structures at the _____ uranium mill facility. This review included an evaluation using the review procedures in Section 5.2.2, and the acceptance criteria outlined in Section 5.2.3 of this standard review plan.

The licensee has provided an acceptable site decommissioning plan, including (1) appropriately substantiated site characterization data or plans to identify contaminated areas; (2) plans to clean up and place within the disposal cell, all materials that are in excess of the standards and approved guidelines, including hazardous material; (3) sufficient information concerning instrumentation and procedures; (4) plans for post-reclamation survey and sampling for verification that the soil and structures meet radiological limits; (5) location for retention of records important to decommissioning; (6) methods to protect workers, the public, and the environment; and (7) a cost estimate for all proposed decommissioning activities.

On the basis of the information presented in the reclamation plan and the detailed review conducted of proposed decommissioning activities for the _____ uranium mill facility, the staff concludes that the information is acceptable and is in compliance with 10 CFR Part 40, Appendix A, Criterion 6(6), which requires that any portion of a licensed uranium mill site not designed to control radon releases, contain a concentration of radium in land, averaged over areas of 100 square meters, which, as a result of byproduct material, does not exceed the background levels by more than (i) 5 pCi/g of Ra-226 averaged over the first 15 cm [6 in.] below the surface, and (ii) 15 pCi/g of Ra-226 averaged over 15-cm-thick layers more than 15 cm below the surface. Also, the cleanup of other residual radionuclides in soil and residual surface activity on structures to remain on site, meet the criteria developed with the radium benchmark dose approach, including a demonstration of "as low as is reasonably achievable" radiation levels and application of the unity test where applicable. For cases in which the licensee has proposed an alternative to the requirements of Criterion 6(6) or the approved guidance, the staff determines that the resulting level of protection is equivalent to that required by this criterion. In addition, the plan demonstrates compliance with 10 CFR Part 40, Appendix A, Criterion 6(7), which requires prevention of threats to human health and the environment from non-radiological hazards associated with the wastes.

The decommissioning plan specifies the location of records of information important to the decommissioning as required by 10 CFR 40.36(f) and meets the criteria of 10 CFR 40.42(g)(4) and (5). The plan sufficiently demonstrates that the proposed decommissioning activities will result in compliance with 10 CFR 40.42(j)(2) requirements to conduct a radiation survey. The plan complies with the 10 CFR 40.42(k)(1) and (2) requirements that source material be properly disposed of and reasonable effort be made to eliminate residual radioactive contamination. The decommissioning cost estimate meets the requirements of 10 CFR 40.42(g)(4)(v) and Appendix A, Criterion 9.

Radiation Protection

5.2.5   References

American National Standards Institute Standards:

N42.17A–1989, "Performance Specifications for Health Physics Instrumentation-Portable Instrumentation for Use in Normal Environmental Conditions."

N42.12–1994, "American National Standard Calibration and Usage of Thallium-Activated Sodium Iodide Detector Systems for Assay of Radionuclides."

American Society for Testing and Materials Standards:

C 998-90 (reaffirmed 1995), "Standard Practice for Sampling Surface Soil for Radionuclides."

D 5283-92, "Standard Practice for Generation of Environmental Data Related to Waste Management Activities: Quality Assurance and Quality Control Planning and Implementation."

E 181-93, "Standard Test Methods for Detector Calibration and Analysis of Radionuclides."

E 1893-97, "Standard Guide for Selection and Use of Portable Survey Instruments for Performing *In Situ* Radiological Assessments in Support of Decommissioning."

National Council on Radiation Protection and Measurements. "Calibration of Survey Instruments Used in Radiation Protection for the Assessment of Ionizing Radiation Fields and Radioactive Surface Contamination." Report No. 112. 1991.

NRC. "Uranium Mill *In-Situ* Leach Uranium Recovery, and 11e(2) byproduct Material Disposal Site Decommissioning Inspection." Inspection Manual—Inspection Procedure 87654. Washington DC: NRC. March 2002.

———. NUREG–1575, "Multi-Agency Radiation Survey and Site Investigation Manual." Rev. 1. Washington DC: NRC. 2000a.

———. NUREG–1727, "NMSS Decommissioning Standard Review Plan." Washington, DC: NRC. 2000b.

———. NUREG–1505, "A Nonparametric Statistical Methodology for the Design and Analysis of Final Status Decommissioning Surveys." Rev. 1. Washington DC: NRC. 1998a.

———. NUREG–1507, "Minimum Detectable Concentrations With Typical Radiation Survey Instruments for Various Contaminants and Field Conditions." Washington DC: NRC. 1998b.

———. "Annual Financial Surety Update Requirements for Uranium Recovery Licensees." Generic Letter 97-03. Washington, DC: NRC. July 9, 1997.

——. Regulatory Guide 3.65, "Standard Format and Content of Decommissioning Plans Under 10 CFR Parts 30, 40, and 70." Washington DC: NRC, Office of Standards Development. 1989.

——. "Technical Position on Financial Assurances for Restoration, Decommissioning, and Long-Term Surveillance and Control of Uranium Recovery Facilities." Washington, DC: NRC. 1988.

5.3 Radiation Safety Controls and Monitoring

5.3.1 Areas of Review

The areas of review for radiation safety for protecting the site worker, the public, and the environment during reclamation and decommissioning are the control of releases, the radiation exposure and environmental monitoring programs, and the contamination control program. Decommissioning activities at mill sites involve occupational, and possibly public, exposures to radioactive materials that may require different or additional monitoring and control procedures than during site operation or standby status. Potential sources of exposure from working with tailings material are caused by airborne particulate contamination, radon gas, and external gamma radiation. Surface activity or dust on equipment and structures to be dismantled or decontaminated could also be a source of exposure.

The reclamation and final decommissioning plans should contain the licensee's evaluation of the site current radiation safety/protection plan or program and any proposed changes to the program for reclamation and decommissioning operations. The proposed measures should keep exposures as low as is reasonably achievable and in compliance with the requirements of 10 CFR Part 20. Key components of the program should address hazards unique to the reclamation or decommissioning work environment. Any new activities that could increase hazards to general health and safety (e.g., cleanup in confined spaces, or removal of hazardous or flammable chemicals) should be identified, considering the NRC Memorandum of Understanding with the Occupational Safety and Health Administration.

5.3.2 Review Procedures

Determine that the proposed radiation safety controls and all monitoring programs and procedures are sufficient to comply with the regulatory requirements during decommissioning and reclamation. A licensee will already have an approved radiation safety program in place; therefore, the focus of the review should be to ensure that the reclamation plan addresses those aspects of worker and public protection that require special consideration in planning reclamation and decommissioning activities. The environmental impacts of these activities will be addressed in the environmental assessment, but any concerns requiring mitigation should be addressed in the reclamation plan. The reclamation plan should confirm the applicability of the radiation protection and monitoring programs to reclamation and decommissioning activities or should propose changes to address new program needs based on review of the following:

Radiation Protection

(1)   Control of Releases

Determine whether the proposed systems and procedures (e.g., tailings stabilization, dust control) are sufficient to minimize environmental emissions from the tailings impoundment construction activities or structure demolition, taking into consideration important release mechanisms such as wind resuspension and surface erosion. Radon gas emanating from the tailings pile is also a radiation safety concern for workers and downwind off-site populations. However, because control of the source is not possible during tailings recontouring or cleanup, the reviewer should examine the means proposed to limit the worker inhalation hazard (i.e., limiting exposure time, or using dust masks or respirators if required) and to establish an acceptable environmental monitoring program for measuring off-site airborne concentrations. Also, liquid releases can be created by rainwater runoff. Therefore, the review of the reclamation plan should include an evaluation, taking all exposure pathways into account, of proposals for ensuring off-site exposures are as low as is reasonably achievable.

(2)   External Radiation Exposure Monitoring Program

Determine if changes to the existing program are needed or if proposed changes are adequate. The reviewer should consider the types of surveys conducted, criteria for determining survey locations, frequency of surveys, action levels, management audits, and corrective action requirements. Also, consider if changes are required in the program for personal/personnel monitoring (dosimeters and air samplers), including the criteria for placing workers in the program.

(3)   Airborne Radiation Monitoring Program for Work Areas

Evaluate whether the proposed sampling locations, frequencies, procedures, and equipment are adequate to determine concentrations of airborne radioactive materials (including radon) in work areas during construction, demolition, and cleanup activities. Action levels, audits, and corrective action requirements should also be evaluated.

(4)   Bioassay Program

Review the existing bioassay program or proposed changes to determine whether the proposed bioassay program is sufficient to protect employees performing decommissioning activities in yellowcake processing areas.

(5)   Contamination Control Program

Evaluate the occupational radiation survey program. This review should include proposed housekeeping and cleanup requirements and specifications for clean areas to control contamination. Action levels for clean areas and for the release of materials, equipment, and work clothes from clean areas and/or the site should be evaluated.

(6)   Environmental Monitoring Program

Determine whether the environmental monitoring program proposed for measuring concentrations and quantities of both radioactive and non-radioactive materials released to, and in, the environs of the proposed facility, are sufficient to protect employees and the public. Potential releases during disposal cell construction and cleanup activities will be primarily from resuspended tailings material and radon gas. The reviewer should focus on the frequency of sampling and analysis, the types and sensitivity of analyses, action levels, corrective action requirements, and the required number of effluent and environmental monitoring stations (including criteria for determining monitor station locations considering the reclamation work). The guidance in Regulatory Guide 4.20 (NRC, 1996) should be considered.

(7)   Record Keeping

Determine whether the record keeping requirements for the radiation protection program have been addressed; that is, records of the provisions of the program and audits or other reviews of content and implementation are maintained for at least 3 years. Other records are maintained according to Subpart L of 10 CFR Part 20 and 10 CFR Part 40, Appendix A, Criterion 6(4).

5.3.3   Acceptance Criteria

The radiation safety controls and monitoring for site worker, public, and environmental protection during reclamation and decommissioning will be acceptable if they meet the following criteria:

(1)   The reclamation plan identifies the radiation safety concerns that are unique to reclamation and decommissioning activities. These concerns include characterization of radiation hazards associated with inhalation of resuspended tailings material or yellowcake, gamma exposure from working close to tailings, and inhalation of radon gas and its progeny (decay products) emanating from tailings material.

(2)   The reclamation plan describes any changes to an existing radiation safety or monitoring program that would be necessary to ensure worker or public safety during reclamation or decommissioning activities.

(3)   Standard dust control measures such as regular wetting and/or phased stabilization are to be used for control of windblown tailings material or yellowcake dust.

(4)   Any proposed changes to the established bioassay program will meet criteria of the applicable parts of Regulatory Guide 8.22, "Bioassay at Uranium Mills" (NRC, 1988) and Regulatory Guide 8.9, Revision 1, "Acceptable Concepts, Models, Equations, and Assumptions for a Bioassay Program" (NRC, 1993), or an acceptable justification is provided for selecting an alternate approach.

(5)   Any proposed workplace airborne radiological monitoring program will support the proposed bioassay program and is consistent with applicable parts of Regulatory Guide 8.25, "Air Sampling in the Workplace" (NRC, 1992) and Regulatory Guide 8.30, "Health

Physics Surveys in Uranium Mills" (NRC, 2002), or an acceptable justification is provided for selecting an alternate approach. The monitoring program will provide adequate protection of workers from radon gas or particulate exposures to maintain compliance with the inhalation limits in 10 CFR Part 20. If sampling locations will be revised, the reclamation plan contains one or more maps of the site that indicate the location of samplers for airborne radiation and provide the justification for determining the revised locations.

(6)   Any proposed contamination control program is consistent with the guidance on conducting surveys for contamination of skin and of personal clothing presented in Regulatory Guide 8.30 (NRC, 2002).

(7)   Any proposed environmental radiological monitoring program is consistent with applicable parts of Regulatory Guide 4.14, "Radiological Effluent and Environmental Monitoring at Uranium Mills" (NRC, 1980), or an acceptable justification is provided for selecting an alternate approach. The licensee has adequately considered site-specific aspects of climate and topography in determining locations of off-site airborne monitoring stations and environmental sampling areas so that detection of maximum off-site concentrations of windblown tailings material and contamination from any other significant transport pathways applicable to the site is ensured.

(8)   Any proposed radiation protection program contains plans for documentation of exposures to all monitored workers and contractors and for availability of exposure records in a single location for inspection. The program provides for recordkeeping that meets the requirements of 10 CFR 20.2102; at least annual review of the program content and implementation; and implementation of the "as low as is reasonably achievable" requirements of 20.1101(d).

5.3.4   Evaluation Findings

If the staff review, as described in this section, results in the acceptance of the radiation safety controls and monitoring for site worker, public, and environmental protection during disposal cell construction and site cleanup, the following conclusions may be presented in the technical evaluation report:

The staff has completed its review of the radiation safety controls and monitoring for site worker, public, and environmental protection during reclamation and decommissioning at the _____ uranium mill facility. This review included an evaluation using the review procedures in Section 5.3.2 and the acceptance criteria outlined in Section 5.3.3 of this standard review plan.

The licensee has provided an acceptable evaluation of radiation safety controls and monitoring required for worker, public, and environmental protection during reclamation and decommissioning activities, including (1) identification of the radiation safety concerns that are unique or likely to increase during reclamation construction and site decommissioning; (2) any necessary changes and associated justifications in the radiation safety program, such as personnel and environmental monitoring; (3) identification and discussion of any changes in an existing radiation protection program; (4) control of potential contamination from windblown tailings by regular wetting and/or phased stabilization; and (5) the monitoring and contamination control programs will allow compliance with applicable portions of 10 CFR Parts 20 and 40.

On the basis of the information presented in the reclamation plan and the detailed review conducted of the radiation safety controls and monitoring for worker, public, and environment protection during reclamation and decommissioning for the _____ uranium mill facility, the NRC staff concludes that the information is acceptable and is in compliance with 10 CFR 20.1101, which requires development, documentation, and implementation of a radiation protection program ensuring compliance with 10 CFR Part 20 requirements and the use of procedures and engineering controls to achieve occupational and public doses that are as low as is reasonably achievable. The 10 CFR Part 40, Appendix A, Criterion 8, requirements for implementation of control measures to limit dust emissions from tailings that are not covered by standing liquids, including wetting or chemical stabilization, will be met. [This requirement may be relaxed for tailings impoundments that have surfaces that are sheltered from wind exposure (i.e., below grade) or that have an interim cover.] The requirements in 10 CFR 40.42(g)(4)(iii), to describe methods that ensure protection of workers and the environment against radiation hazards during decommissioning, have been met.

5.3.5 References

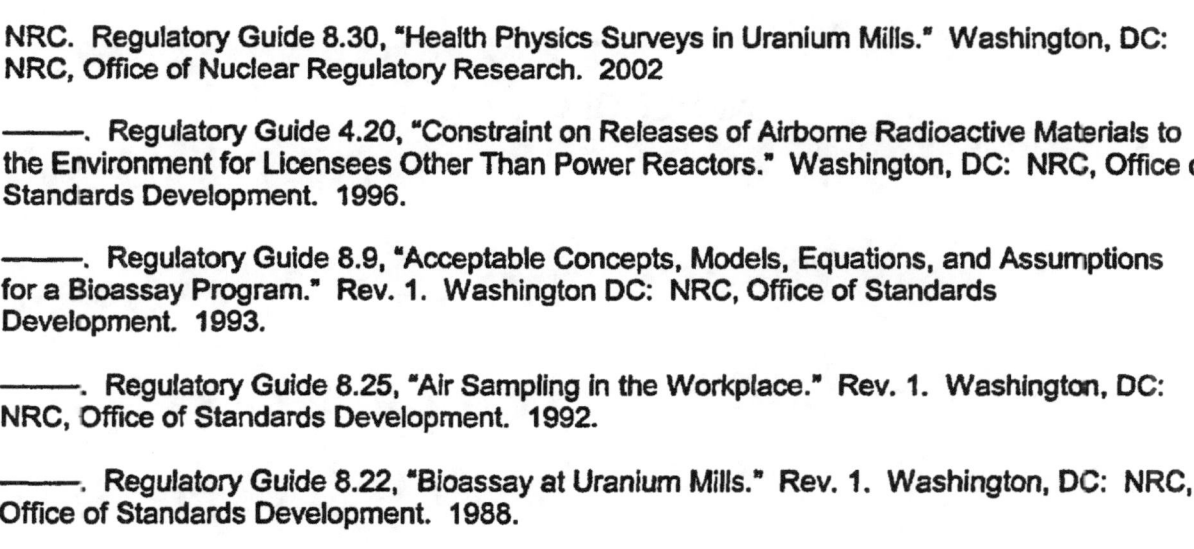

NRC. Regulatory Guide 8.30, "Health Physics Surveys in Uranium Mills." Washington, DC: NRC, Office of Nuclear Regulatory Research. 2002

———. Regulatory Guide 4.20, "Constraint on Releases of Airborne Radioactive Materials to the Environment for Licensees Other Than Power Reactors." Washington, DC: NRC, Office of Standards Development. 1996.

———. Regulatory Guide 8.9, "Acceptable Concepts, Models, Equations, and Assumptions for a Bioassay Program." Rev. 1. Washington DC: NRC, Office of Standards Development. 1993.

———. Regulatory Guide 8.25, "Air Sampling in the Workplace." Rev. 1. Washington, DC: NRC, Office of Standards Development. 1992.

———. Regulatory Guide 8.22, "Bioassay at Uranium Mills." Rev. 1. Washington, DC: NRC, Office of Standards Development. 1988.

———. Regulatory Guide 4.14, "Radiological Effluent and Environmental Monitoring at Uranium Mills." Rev. 1. Washington, DC: NRC, Office of Standards Development. 1980.

APPENDIX A

RELATIONSHIP OF 10 CFR PART 40, APPENDIX A REQUIREMENTS TO STANDARD REVIEW PLAN SECTIONS

This appendix identifies the specific standard review plan sections where the criteria of 10 CFR Part 40, Appendix A are addressed.

10 CFR Part 40, Appendix A Criterion	Locations in NUREG–1620 Where the Criterion is Addressed
Criterion 1: Optimize site selection to achieve permanent isolation of tailings without maintenance.	2.1.4, 3.1.4, 3.2.4, 3.3.4, 3.4.4, 3.5.4
Criterion 2: Avoid proliferation of small waste disposal sites.	Not applicable to this standard review plan.
Criterion 3: Dispose of tailings below grade or provide equivalent isolation.	2.1.4
Criterion 4: Adhere to siting and design criteria.	
(a) Minimize upstream rainfall catchment areas.	3.2.4
(b) Select topographic features that provide good wind protection.	3.5.4
(c) Provide relatively flat embankment and cover slopes.	2.2.4, 2.4.4, 2.5.4, 2.6.4, 2.7.4, 3.4.4, 3.5.4
(d) Establish a self-sustaining vegetative cover or rock cover considering stability, erosion potential, and geomorphology.	2.2.4, 2.6.4, 3.4.4
(e) Locate away from faults capable of causing impoundment failure.	1.1.4, 1.2.4, 1.4.4, 2.1.4, 2.2.4
(f) Design to promote deposition, where feasible.	3.4.4
Criterion 5A: Meet the primary ground-water protection standard.	
(1) Design, construct, and install an impoundment liner that prevents migration of wastes to subsurface soil, ground water, or surface water.	Not applicable to this standard review plan.
(2) Construct liner of suitable materials, place it on an adequate base, and install it to cover surrounding earth likely to be in contact with wastes or leachate.	Not applicable to this standard review plan.

10 CFR Part 40, Appendix A Criterion	Locations in NUREG–1620 Where the Criterion is Addressed
(3) Apply alternate design or operating practices that will prevent migration of hazardous constituents into ground water or surface water.	Not applicable to this standard review plan.
(4) Design, construct, maintain, and operate impoundments to prevent overtopping.	4.4.4
(5) Design, construct, and maintain dikes to prevent massive failure.	2.2.4, 4.4.4
Criterion 5B: Conform to the secondary groundwater protection standards.	
(1) Prevent hazardous constituents from exceeding specified concentration limits in the uppermost aquifer beyond the point of compliance.	4.1.4, 4.2.4, 4.3.4, 4.4.4
(2) Define hazardous constituents as those expected to be in or derived from the byproduct material, those detected in the uppermost aquifer, and those listed in Criterion 13.	4.1.4, 4.2.4, 4.3.4, 4.4.4
(3) Exclude hazardous constituents if they are not capable of posing a substantial present or potential hazard to human health or the environment.	4.1.4, 4.2.4, 4.3.4, 4.4.4
(4) Consider identification of underground sources of drinking water and exempted aquifers.	4.1.4, 4.2.4, 4.3.4, 4.4.4
(5) Ensure hazardous constitutents at the point of compliance do not exceed the background concentration, the value in Paragraph 5C, or an approved alternate concentration limit.	4.1.4, 4.2.4, 4.3.4, 4.4.4
(6) Establish alternate concentration limits, if necessary, after considering practical corrective actions, as low as is reasonably achievable requirements, and potential hazard to human health or the environment.	4.1.4, 4.2.4, 4.3.4, 4.4.4
Criterion 5C: Comply with maximum values for ground-water protection.	4.1.4, 4.2.4, 4.3.4, 4.4.4

10 CFR Part 40, Appendix A Criterion	Locations in NUREG–1620 Where the Criterion is Addressed
Criterion 5D: Implement a ground-water corrective action program if secondary ground water protection standards are exceeded.	4.4.4
Criterion 5E: Consider appropriate measures when developing and conducting a ground-water protection program.	
(1) Incorporate leak detection systems for synthetic liners and conduct appropriate testing for clay/soil liners.	4.1.4, 4.4.4
(2) Use process designs that maximize solution recycling and water conservation.	4.1.4, 4.4.4
(3) Dewater tailings by process devices or properly designed and installed drainage systems.	4.1.4, 4.4.4
(4) Neutralize hazardous constituents to promote immobilization.	4.1.4, 4.4.4
Criterion 5F: Alleviate seepage impacts where they are occurring and restore ground-water quality.	4.1.4, 4.3.4, 4.4.4
Criterion 5G: Provide appropriate information for a tailings disposal system.	
(1) Define the chemical and radioactive characteristics of waste solutions.	4.1.4, 4.3.4, 4.4.4
(2) Describe the characteristics of the underlying soil and geologic formations.	1.1.4, 2.1.4, 4.1.4, 4.3.4, 4.4.4
(3) Define the location, extent, quality, capacity, and current uses of ground water.	4.1.4, 4.3.4, 4.4.4
Criterion 5H: Minimize penetration of radionuclides into underlying soils when stockpiling.	4.1.4, 4.4.4
Criterion 6: Install an appropriate cover and close the waste disposal area.	
(1) Ensure the cover meets lifetime and radon flux	1.1.4, 1.2.4, 1.3.4, 1.4.4, 2.1.4,

10 CFR Part 40, Appendix A Criterion	Locations in NUREG–1620 Where the Criterion is Addressed
release specifications.	2.2.4, 2.3.4, 2.4.4, 2.5.4, 2.6.4, 2.7.4, 3.2.4, 3.3.4, 3.4.4, 3.5.4, 4.3.4, 5.1.4
(2) Demonstrate the effectiveness of the final radon barrier prior to placement of erosion protection barriers or other features.	5.1.2.1(b)
(3) Demonstrate the effectiveness of phased emplacement of radon barriers as each section is completed, if applicable.	5.1.4
(4) Document verification of radon barrier effectiveness to the U.S. Nuclear Regulatory Commission (NRC) and maintain records of this verification.	2.7.4
(5) Ensure that radon exhalation is not significantly above background because of the cover material.	5.1.4
(6) Clean up residual contamination from byproduct material consistent with the radium benchmark dose.	5.2.4
(7) Prevent threats to human health and the environment from non-radiological hazards.	5.2.4
Criterion 6A: Ensure expeditious completion of the final radon barrier.	
(1) Complete the radon barrier as expeditiously as practical after ceasing operations in accordance with a written, Commission-approved reclamation plan.	2.6.4, 5.2.4
(2) Extend milestone completion dates if justified by radon release levels, cost considerations consistent with available technology.	Requirement on Commission.
(3) Authorize disposal of byproduct materials or similar materials from other sources if appropriate criteria are met.	Requirement on Commission.

10 CFR Part 40, Appendix A Criterion		Locations in NUREG-1620 Where the Criterion is Addressed
Criterion 7:	Conduct pre-operational and operational monitoring programs.	4.1.4
Criterion 7A:	Establish a detection monitoring program to set site-specific ground-water protection standards, a compliance monitoring system once ground-water protection standards have been established, and a corrective action monitoring program in conjunction with a corrective action program.	4.1.4, 4.2.4, 4.3.4, 4.4.4
Criterion 8:	Conduct milling operations, including ore storage, tailings placement, and yellowcake drying and packaging operations so that airborne releases are as low as is reasonably achievable .	Not applicable to this standard review plan.
Criterion 8A:	Conduct and record daily inspections of tailings or waste retention systems and report failures or unusual conditions to NRC.	4.4.4
Criterion 9:	Establish appropriate financial surety arrangements for decontamination, decommissioning, and reclamation.	4.4.4, 5.2.4
Criterion 10:	Establish sufficient funds to cover the costs of long-term surveillance and control.	4.4.4
Criterion 11A:	Comply with effective date for site and byproduct material ownership requirements.	Requirement on Commission.
Criterion 11B:	Establish license conditions or terms to ensure that licensees comply with ownership requirements prior to license termination for sites used for tailings disposal.	Requirement on Commission.
Criterion 11C:	Transfer title to byproduct material and land to the United States or the state in which the land is located.	Not applicable to this standard review plan.

10 CFR Part 40, Appendix A Criterion		Locations in NUREG–1620 Where the Criterion is Addressed
Criterion 11D:	Permit use of surface and subsurface estates if the public health, safety, welfare, or environment will not be endangered.	Requirement on Commission.
Criterion 11E:	Transfer material and land to the United States or a State without cost other than administrative legal costs.	Not applicable to this standard review plan.
Criterion 11F:	Follow specific requirements for land held in trust for or owned by Indian Tribes.	Requirement on Commission.
Criterion 12:	Minimize or avoid long-term active maintenance and conduct and report on annual inspections.	3.2.4, 3.3.4, 3.4.4, 3.5.4
Criterion 13:	Establish standards for constituents reasonably expected to be in or derived from byproduct materials and detected in ground water.	4.1.4, 4.2.4, 4.4.4

APPENDIX B

GUIDANCE TO THE U.S. NUCLEAR REGULATORY COMMISSION STAFF FOR REVIEWING HISTORICAL ASPECTS OF SITE PERFORMANCE FOR LICENSE RENEWALS AND AMENDMENTS

For license renewals and amendments, the historical record of site operations contains valuable information for evaluating the licensing actions. Following are specific areas in which a compliance history or record of site operations and changes should be presented for review:

- Amendments and changes to operating practices or procedures.

- License violations identified during U.S. Nuclear Regulatory Commission (NRC) or Agreement State site inspections.

- Cleanup histories or status.

- Exceedances of any regulatory standard or license condition pertaining to radiation exposure, contamination, or release limits.

- Exceedances of any regulatory standard or licenses condition pertaining to non-radiation contaminant exposure or release limits.

- Changes to any site characterization information important to the evaluation of the reclamation plan, such as changes to site location and layout, uses of adjacent lands and waters, meteorology, seismology, the geologic or hydrologic setting, ecology, background radiological or non-radiological characteristics, and other environmental features.

- Results of site operations such as data on radiological and non-radiological effects, accidents, and the economic effects of operations.

- Changes to factors that may cause reconsideration of alternatives to the proposed action.

- Changes to the economic costs and benefits for the facility since the last application.

If, after reviewing these historical aspects of site operations, the staff concludes that the site has been operated so as to protect health, safety, and the environment, and that no unreviewed safety-related concerns have been identified, only those changes proposed by the license renewal or amendment or application should be reviewed, using the appropriate sections of this standard review plan. Aspects of the facility and its operations that have not changed since the last license renewal or amendment should not be reexamined.

APPENDIX C

OUTLINE RECOMMENDED BY THE U.S. NUCLEAR REGULATORY COMMISSION STAFF FOR PREPARING SITE-SPECIFIC FACILITY RECLAMATION AND STABILIZATION COST ESTIMATES FOR REVIEW

As required by Criteria 9 and 10 of 10 CFR Part 40, Appendix A, the licensee shall supply sufficient information for the U.S. Nuclear Regulatory Commission (NRC) to verify that the amount of coverage provided by the financial assurance will permit the completion of all decontamination, decommissioning, and reclamation of sites, structures, and equipment used in conjunction with byproduct material. Cost estimates for the following items (where applicable) should be submitted to NRC with the initial license application or reclamation plan, and should be updated annually as specified in the license. Cost estimates must be calculated on the basis of completion of all activities by a third party. Unit costs, calculations, references, assumptions, equipment and operator efficiencies, *et cetera*, must be provided. The annual surety estimate must be prospective of all work to be performed at the site. The licensee must provide estimated costs for all decommissioning, reclamation, and ground-water cleanup work remaining to be performed at the site, not simply deduct the cost of work already performed from the previous surety estimate [see NRC Generic Letter 97-03 (NRC, 1997)]. The licensee can propose to deduct for work done and approved by NRC as meeting specifications.

The detailed cost information necessary to verify the cost estimates for the preceding categories of closure work is summarized in the recommended outline that follows. For each area, estimates should the include costs for equipment; materials; labor and overhead; licenses, permits, and miscellaneous site-specific costs; and any other activity or resource that will require expenditure of funds.

(I)   Facility Decommissioning

This includes dismantling and decontamination, or disposal of all structures and equipment. This work may be done in two phases. In the first phase, only the equipment not used for ground-water cleanup (including the stability monitoring period) is removed. Removal of the remaining equipment would be performed in a second phase, after the approved completion of ground-water cleanup. The buildings may be decontaminated and released for unrestricted use.

(A)   Salvageable building and equipment decontamination. For each building or piece of equipment listed, the following data should be provided:

(1)   Area of contamination

(2)   Survey costs

(3)   Decontamination costs

(B)   Non-salvageable building and equipment demolition and disposal:

(1)   List of major categories of building and equipment to be disposed of and their corresponding quantities:

        (a)   Structures (list each major), metric tons [tons(short)] of material, and building volume cubic meters (cubic yards)

        (b)   Foundation concrete [cubic meters (cubic yards)]

        (c)   Process equipment [metric tons (tons (short)]

        (d)   Piping and insulation (lump sum)

        (e)   Electrical and instrumentation (lump sum)

    (2)   Disposal of chemical solutions within the facility.

(C)   Cleanup of contaminated areas (ore storage pad, access roads, process area, evaporation pond residues, etc.).

Reclamation—This entails recontouring the tailings disposal cell and evaporation ponds and placing top soil or other materials acceptable to NRC. Reclamation may also include cleanup of windblown materials and revegetation, including:

    (1)   Cleanup of windblown materials (e.g., volume and area, unit cost/cubic yard).

    (2)   Placement of borrow materials removal (e.g., rental rate, cost/cubic yard).

    (3)   Dust suppression and site maintenance.

    (4)   Monitoring and testing of construction.

    (5)   Regrading.

    (6)   Placement of the frost barrier.

    (7)   Placement of the radon barrier.

    (8)   Installation of erosion protection and armor.

    (9)   Installation of any vegetative cover.

    (10)   Design and construction of drainage ditches.

    (11)   Recontouring of land surfaces.

    (12)   Revegetation.

(II)   Ground-Water Cleanup and Well Decommissioning

Ground-water cleanup is done in accordance with an approved corrective action plan. The costs include water treatment equipment, operation, maintenance, and component replacement.

    (A)   Method of cleanup

    (B)   Volume of aquifer required to be restored, area and thickness of aquifer, number of required pumping cycles, and cycling time

    (C)   Verification sample analysis

    (D)   Well decommissioning:

        (1)   Number of drill holes to be plugged

        (2)   Depth and size of each drill hole

        (3)   Material to be used for plugging including acquisition, transportation, and plugging

(III)   Radiological Survey and Monitoring

Radiological Survey—Surveys and soil samples for radium in areas to be released for restricted use. Soils around the tailings disposal cell, evaporation ponds, and process buildings should be analyzed for radium content. A gamma survey of all areas should be made before release for unrestricted use. All equipment released for unrestricted use should be surveyed and records maintained.

    (A)   Soil samples for radium (and uranium and thorium, if needed) (e.g., number, cost to collect, and analyze)

    (B)   Decommissioning equipment and building smear samples and alpha surface surveys

    (C)   Gamma survey frequency, location, and techniques (e.g., type, number, unit cost)

    (D)   Environmental monitoring

    (E)   Personnel monitoring

Appendix C

(IV)   Project Management Costs and Miscellaneous

Itemize estimated costs associated with project management; engineering design, review, and change; mobilization; legal expenses; power during reclamation; quality control; radiological safety; and any costs not included in other estimation categories. costs should include preparation of completion report and license termination activities.

Potential needs for future well maintenance or replacement are identified. If periodic well replacement is projected, an increase in the long-term care payment is included (American Society for Testing and Materials Standard D 5978).

(V)   Labor and Equipment Overhead, Contractor Profit

Overhead costs for labor and equipment and contractor profit may be calculated as separate items or loaded into hourly rates. If included in hourly rates, the unit costs must identify the percentages applied for each area.

(VI)   Long-term Surveillance Fee

The fee required by 10 CFR Part 40, Appendix A, Criterion 10, to include cost of any required long-term monitoring (e.g., ground water) or maintenance (e.g., fences, vegetation control).

(VII)   Contingency

The licensee should add a contingency amount to the total cost estimate for the final site closure. The staff currently considers a 15 percent contingency to be an acceptable minimum amount.

(VIII)   Adjustments to Surety Amounts

The licensee is required by 10 CFR Part 40, Appendix A, Criterion 9, to adjust cost estimates annually to account for inflation and changes in reclamation plans. The submittal should be in the form of a request for amendment to the license.

(A)   Adjustments for inflation: The licensee should submit a revised surety incorporating adjustments to the cost estimates for inflation 90 days before each anniversary of the date on which the first reclamation plan and cost estimate was approved. The adjustment should be made using the inflation rule indicated by the change in the Urban Consumer Price Index published by the U.S. Department of Labor, Bureau of Labor Statistics.

(B)   Changes in Plans:

(1)   Changes in the process, such as size or method of operation.

(2)	Licensee-initiated changes in reclamation plans or reclamation/ decommissioning activities performed.

(3)	Adjustments to reclamation plans required by NRC.

(4)	Proposed revisions to reclamation plans must be thoroughly documented and cost estimates and the basis for cost estimates must be detailed for NRC review and approval.

To avoid unnecessary duplication and expense, NRC shall take into account surety arrangements required by other federal agencies, state agencies, or other local governing bodies. However, the Commission is not required to accept such sureties if they are not sufficient. Similarly, no reduction to surety amounts established with other agencies shall be effected without NRC approval. Copies of all correspondence relating to the surety between the licensee and the state should be submitted to NRC. If authorized by NRC to maintain a surety with the state as the beneficiary, it is the responsibility of the licensee to give NRC verification of that surety; ensure that the agreement with the State specifically identifies the financial surety's application, uranium mill tailings site, and decommissioning/reclamation requirements; and transfer the long-term surveillance and control fee to the U.S. Department of the Treasury before license termination.

All costs (unit and total) are to be estimated on the basis of third party independent contractor costs (include overhead and profit in unit costs or as a percentage of the total). Equipment owned by the licensee and the availability of licensee staff should not be considered in the estimate to reduce cost calculations. All costs should be based on current-year dollars. Credit for salvage value is generally not acceptable on the estimated costs.

NRC staff review may include a comparison of unit cost estimates with standard construction cost guides (e.g., R.S. Means, Dodge Guide, Data Quest) and discussions with appropriate state or local authorities (e.g., highway cost construction). The licensee should provide supporting information or the basis for selection of the unit cost figures used in estimates. The staff may elect to use a publicly available computer code such as RACER™ (Talisman Partners, Ltd., 2000) or spreadsheet to assess these costs.

References

American Society for Testing and Materials Standards

D 5978, "Standard Guide for Maintenance and Rehabilitation of Ground-water Monitoring Wells."

NRC. "Annual Financial Surety Update Requirements for Uranium Recovery Licensees." Generic Letter 97-03. Washington, DC: NRC. July 1997.

Talisman Partners, Ltd. "Introduction to RACER 2000™ (Version 2.1.0)—A Quick Reference." Englewood, Colorado: Talisman Partners, Ltd. 2000.

APPENDIX D

GUIDANCE TO THE U.S. NUCLEAR REGULATORY COMMISSION STAFF FOR REVIEWING LONG-TERM SURVEILLANCE PLANS

D1.0 BACKGROUND

The Atomic Energy Act of 1954, as amended (hereafter, the Act), contains the statutory requirements for transfer of the title and custody of byproduct material and any land used for the disposal of such byproduct material from a uranium mill licensee to either federal or state control, before termination of the licensee-specific license. These requirements are codified in 10 CFR Part 40, at Section 40.28, "General license for custody and long-term care of uranium or thorium byproduct materials disposal sites." Section 40.28, along with pertinent requirements stated in Appendix A to 10 CFR Part 40 (hereafter Appendix A), requires the completion of certain licensing actions before the transfer of the land and byproduct material to the United States or the appropriate state for long-term care. As part of the license termination process, the intended custodial agency, federal or state government, will prepare a long-term surveillance plan for review and concurrence/acceptance by the U.S. Nuclear Regulatory Commission (NRC). The long-term surveillance plan must document the general licensee's plan for long-term care, including inspection, monitoring, maintenance, and emergency measures necessary to protect public health and safety. This document presents guidance to the NRC staff on review of the long-term surveillance plan. Standard review plan Appendix E presents guidance on the license termination process, and presents the role of the long-term surveillance plan in the overall context of the license termination process.

Review and acceptance of long-term surveillance plans is the sole responsibility of the NRC. However, Agreement State comments prepared using this guidance are welcomed and will be considered, if provided.

D2.0 REVIEW OF LONG-TERM SURVEILLANCE PLAN

D2.1 Areas of Review

In accordance with 10 CFR 40.28(b), the long-term surveillance plan should present the following information:

(1)   A legal description of the disposal site to be transferred and licensed

(2)   A detailed description of the final conditions of the disposal site, including existing ground-water characterization

(3)   A description of the long-term surveillance program, including proposed inspection frequency and reporting to the Commission; frequency and extent of ground-water monitoring, if required; appropriate constituent limits for ground water; inspection personnel qualifications; inspection procedures; record keeping; and quality assurance procedures

(4)   The criteria for followup inspections in response to unusual observations from routine inspections or extreme natural events

Appendix D

(5)   The criteria for instituting maintenance or emergency measures

D2.2   Information Reviewed

D2.2.1 Legal Description and Ownership of the Land

The reviewer should examine the documents to ensure that the ownership and legal description of the land are satisfactory. The review should include review of (1) the legal description of the disposal site; (2) a brief narrative of the disposal area land ownership, including the number of acres involved and the type of real estate instruments associated with the acquisitions; (3) information associated with the land transaction [i.e., book, page, county, State, and date of deeds; and agreement number and date associated with DOE/tribe agreement (waiver of liability from tribe when land is part of a reservation or has trust status)]; and (4) a statement that real estate correspondence and instruments are maintained and filed by the property management branch of the pertinent custodial agency. The documentation should clearly establish the custodial agency's land ownership when the land transfer takes place.

D2.2.2 Final Condition of the Disposal Site

The reviewer should examine the following: (1) documentation of defined and characterized final closure site condition; (2) as-built drawings; (3) description of disposal cell design; (4) final topographic maps; (5) vicinity maps; (6) ground and aerial photographs; (6) number, location, and condition of survey monuments, site markers, and signs; and (7) existing ground-water characterization and protection activities (if necessary), ground-water monitoring well network to detect changes in ground-water quality from tailings (including evaluating the monitoring data to quantify the rate and magnitude of change). Some of the information may be referenced to the information already submitted to NRC (such as the completion report), and the staff findings on the previously submitted information may be used in this review. It is noted that the final disposition of the tailings residual radioactive material, or wastes at the milling site, should be such that ongoing active maintenance is not necessary to preserve isolation. The descriptions of the final disposal site condition, the ground-water condition, and the proposed ground-water monitoring program should be of sufficient detail that future inspectors have a baseline to determine changes to the site.

D2.2.3 Long-Term Surveillance Program

The staff should review the surveillance (inspection and monitoring) program for:

(1)   Frequency of Inspection—The physical condition of the site (fence, site markers, drains/ditches, rock-mulch/vegetative cover, etc.) should be inspected annually to determine any need for maintenance or monitoring or both. In addition, an inspection should follow an unusual event, such as a heavy storm or an earthquake. On the basis of a site-specific evaluation, NRC may require more frequent site inspections because of the particular features of a disposal site.
(2)   Reporting to the Commission—Results of the inspections for all the sites under the licensee's jurisdiction will be reported to NRC annually within 90 days of the last site inspection in that calendar year. However, any site at which unusual damage or

D–2

disruption is discovered during the inspection requires a preliminary site inspection report to be submitted within 60 days.

(3)   Ground-Water Monitoring—The reviewer should examine long-term surveillance plans to ensure that a ground-water monitoring program is in place to verify that the ground-water quality at the site will continue to meet applicable standards. In particular, the reviewer should determine whether:

(a)   Background, point of compliance, and, if applicable, point of exposure wells have been located as described in the existing license. Wells should be correctly placed as to surface locations and aquifer completions. Well locations should be surveyed and located on site maps drawn to scale.

(b)   The same ground-water protection standards (point of ground-water protection standards or alternate concentration limits) as in the existing license continue to apply. If there has been no leakage from the impoundment into the ground-water, appropriate ground-water parameters should be monitored and detection concentrations should be established that will give early warning of leakage. Appropriate parameters should be indicative of the tailings material and not significantly affected by retardation reactions. For acid tailings, appropriate detection parameters might include total dissolved solids, chloride, or sulfate.

(c)   The sampling frequency is sufficient to protect the public and environment at the point of exposure and sufficient to ensure that the ground-water downgradient of the point of compliance will not be degraded to any great extent before contamination is detected. This will require a knowledge of potential contaminant plume velocities. It is anticipated that the calculation of potential contaminant plume velocities will be based on advective calculations (e.g., American Society for Testing and Materials Standards D 5447, D 5490, D 5609, D 5610, D 5611, D 5718, E 978; and Anderson and Woessner, 1992). However, more complex calculations that include such processes as dispersion and retardation may be performed if site conditions warrant them. For sites with alternate concentration limits, the sampling frequency should be sufficient to detect a potential contaminant plume, well before ground water at the point of exposure is degraded.

It is anticipated for most sites that routine monitoring once every 3 years will be acceptable unless site-specific conditions warrant an increased or decreased frequency of monitoring. If more frequent monitoring is required, the reviewer should assess the increase in the long-term care payment that must be made to support the more frequent monitoring. This increase should be included in the existing surety in addition to the long-term care payment made at the time of license termination.

(d)   Water quality sampling and analysis procedures use appropriate American Society for Testing and Materials or equivalent standards. Wells are constructed to prevent surface-water contamination and are capped and secured to prevent tampering (American Society for Testing and Materials Standard D 5787).

(e)   Actions that the long-term custodian would take should ground-water protection standards be exceeded are described.

If the staff review results in acceptance of the long-term surveillance plan, the staff may conclude that the DOE will conduct long-term surveillance that will confirm that constituents of concern will remain below the relevant standards in 10 CFR Part 40, Appendix A, Criteria 5B(5) and (6). The staff may also conclude that enough funds are available to cover the costs of long-term surveillance and control as required in 10 CFR Part 40, Appendix A, Criterion 10, and that site inspections are planned as required in 10 CFR Part 40, Appendix A, Criterion 12.

(4)   Inspection Personnel Qualifications—The inspection team should be qualified to inspect such site features as subsidence and cracking; erosion by surface water; degradation of erosion protection (rock mulch cover or vegetative cover); integrity of site markers, fences, and settlement plates; and monitoring to verify the presence and concentration limits of hazardous constituents in the ground water. For inspections that follow unusual events, the team should consist of technical personnel of appropriate disciplines.

(5)   Inspection Procedures—The long-term surveillance plan should present details of the inspection procedures such as checklists of items to be inspected, measurements or observations to be made, procedures for documenting the inspection data (photo, video, aerial photo as needed); and duration of inspection (1 to 2 days).

(6)   Recordkeeping and Quality Assurance Procedures—Inspection data should be retained in a format suitable for future retrieval on a long-term basis. The quality assurance aspect of the collection of site and ground-water data, interpretation of the collected data, report preparation, and long-term retention of data should be reviewed.

D2.2.4 Criteria for Follow-up Inspections

The criteria for followup inspections in response to unusual observations from routine inspections or extreme natural events should be reviewed.

(1)   If any unusual observation from the inspection warrants a detailed evaluation, then an unscheduled inspection (followup inspection) will be conducted for a detailed evaluation of the unusual observation encountered in the earlier inspection. The plan should discuss the level of physical distress to the site (settlement/crack magnitude, extent of subsidence, extent of degradation of erosion protection, etc.) and limits of the constituents not to be exceeded in the ground-water that would warrant a further detailed evaluation of the problem to determine the need for a cleanup activity.

(2)   Occurrence of extreme natural events, such as large-magnitude storms and earthquakes or drought, warrants an inspection to verify the physical condition/integrity of the disposal site. The plan should present the magnitude of the natural events that would trigger this inspection.

D2.2.5 Criteria for Instituting Maintenance or Emergency Measures

The plan should present the criteria or the events that will trigger the initiation of maintenance and other emergency measures to restore the integrity of the disposal site and to protect the health and safety of the public. Quantitative and, if not practical, qualitative criteria that would trigger these measures should be discussed in the long-term surveillance plan.

D3.0 CONCLUSIONS

On the basis of its review of the long-term surveillance plan, the staff should be able to conclude that the long-term surveillance plan is in compliance with (1) the content requirements in 10 CFR 40.28(b), (2) the ownership of site and byproduct material requirement in Criterion 11 of Appendix A, and (3) the surveillance plan requirement in Criterion 12 of Appendix A. If the long-term surveillance plan is in compliance with these requirements, the staff can accept it.

D4.0 REFERENCES

American Society for Testing and Materials Standards

D 5447, "Standard Guide for Application of a Ground-water Flow Model to a Site-Specific Problem."

D 5490, "Standard Guide for Comparing Ground-water Flow Model Simulations to Site-Specific Information."

D 5609, "Standard Guide for Defining Boundary Conditions in Ground-water Flow Modeling."

D 5610, "Standard Guide for Defining Initial Conditions in Ground-water Flow Modeling."

D 5611, "Standard Guide for Conducting a Sensitivity Analysis for a Ground-water Flow Model Application."

D 5718, "Standard Guide for Documenting Ground-water Flow Model Application."

D 5787, "Standard Practice for Monitoring Well Protection."

E 978, "Standard Practice for Evaluating Mathematical Models for the Environmental Fate of Chemicals."

Anderson, M.P. and W.W. Woessner. *Applied Ground-Water Modeling: Simulation of Flow and Transport.* New York, New York: Academic Press. 1992.

APPENDIX E

GUIDANCE TO THE U.S. NUCLEAR REGULATORY COMMISSION STAFF ON THE LICENSE TERMINATION PROCESS FOR LICENSEES OF CONVENTIONAL URANIUM MILLS

E1.0 BACKGROUND

The Atomic Energy Act of 1954, as amended, contains the statutory requirements for the transfer of the title and custody to byproduct material and any land used for the disposal of such byproduct material from a uranium mill licensee to either Federal or State control, before termination of the licensee's specific license. These requirements are codified in 10 CFR Part 40, at Section 40.28, "General license for custody and long-term care of uranium or thorium byproduct materials disposal sites." Section 40.28, along with pertinent requirements stated in Appendix A to 10 CFR Part 40 (hereafter Appendix A), provides for the completion of certain licensing actions before the transfer of the land and byproduct material to the United States or the State where the disposal site is located for long-term care.

This document gives the U.S. Nuclear Regulatory Commission (NRC) staff specific directions to be applied in the course of the license termination process for Uranium Mill Tailings Radiation Control Act of 1978 Title II sites that are under NRC jurisdiction. For the license termination of Title II sites that are under Agreement State jurisdiction, guidance is provided in the Office of State and Tribal Programs SA–900 Procedure (NRC, 2001). The license termination process, including the roles of the respective agencies and organizations involved in this process, is discussed in general. Various relevant issues are addressed in greater detail. This is the initial version of this guidance document, and as specific uranium mill licenses are terminated and title to the land and byproduct material is transferred to the appropriate government agency, future revisions are likely to be necessary. These revisions will address not only issues yet to be identified, but also will provide any additional necessary clarification of issues discussed herein.

E2.0 ROLES OF INVOLVED ORGANIZATIONS

E2.1 NRC

In accordance with Section 83c of the Atomic Energy Act, as amended, NRC determines whether the licensee has met all applicable standards and requirements or whether a licensee-proposed alternative meets the standards. This will involve NRC review of licensee submittals relative to the completion of decommissioning, reclamation, and, if necessary, ground-water cleanup.

In addition, the staff should review the site long-term surveillance plan submitted by the custodial agency, for both NRC and Agreement State sites. On NRC acceptance of the long-term surveillance plan, NRC terminates the specific license and places the long-term care and surveillance of the site by the custodial agency under the general license provided at 10 CFR 40.28.

A final NRC responsibility is the determination of the final amount of long-term site surveillance funding. Criterion 10 of Appendix A specifies a minimum charge of $250,000 (1978 dollars), revised to reflect inflation, which may be escalated on a site-specific basis because of

Appendix E

surveillance and long-term monitoring controls beyond those specified in Criterion 10 of
Appendix A. Detailed discussion of the bases used in developing the minimum charge and any
escalated costs appears in Section E3.4 (below).

E2.2 Uranium Mill Licensees

Before license termination, licensees are required by license conditions to complete site
decontamination and decommissioning and surface and ground-water remedial actions
consistent with decommissioning, reclamation, and ground-water corrective action plans.

Licensees must document the completion of these remedial actions in accordance with
procedures developed by NRC. As discussed in Section E3.1 (below), this information will
include a report documenting completion of tailings disposal cell construction, as well as
radiation surveys and other information required under 10 CFR 40.42.

Because the long-term surveillance plan must reflect the remediated condition of the site, the
licensee will work with the custodial agency in preparing the long-term surveillance plan. Most
likely, this will involve supplying the custodial agency with appropriate documentation
(e.g., as-built drawings) of the remedial actions taken and reaching agreements (formal or
informal) with the custodial agency regarding the necessary surveillance control features of the
site (e.g., boundary markers, fencing). It is the custodial agency responsibility to submit the
long-term surveillance plan to NRC for approval. However, the licensee may elect to help
prepare the long-term surveillance plan, to whatever degree is agreed between the licensee
and the custodial agency.

Finally, the licensee provides the funding to cover long-term surveillance of the site, in
accordance with Criterion 10 of Appendix A. NRC will determine the final amount of this charge
on the basis of final conditions at the site.

After termination of the existing license and transfer of the site and byproduct materials to the
custodial agency, the licensee remaining liability extends solely to any fraudulent or negligent
acts committed before the transfer to the custodial agency, as provided for in Section 83b(6) of
the Atomic Energy Act, as amended.

E2.3 Custodial Agency

Section 83 of the Atomic Energy Act, as amended, states that before termination of the specific
license, title to the site and byproduct materials should be transferred to either (1) the
U.S. Department of Energy (DOE); (2) a Federal agency designated by the President; or (3) the
state in which the site is located, at the option of the State. It is expected that the DOE will be
the custodial agency for most, if not all, of the sites.

It is the responsibility of the custodial agency to submit the long-term surveillance plan to NRC
for review and acceptance. Provisions and activities identified in the final long-term surveillance
plan will form the bases of the custodial agency long-term surveillance at the site. The NRC
general license in 10 CFR 40.28(a) becomes effective when the licensee's current specific
license is terminated and the Commission accepts the long-term surveillance plan. Custodial

agencies are required, under 10 CFR 40.28(c)(1) and (c)(2), to implement the provisions of the long-term surveillance plan. These activities could include not only those reflected in the long-term surveillance plan, but also activities voluntarily committed to by the custodial agency.

E2.4 States

As discussed in Section 2.3 (above), the State in which the disposal site is located has the option of becoming the custodial agency. This "right of first refusal" may be exercised either on a site-by-site basis or generally, covering all sites within the State's limits. This option should be exercised early enough in the license termination process so that termination of the specific license and transfer of the site to the appropriate custodial agency are not delayed unnecessarily. Written confirmation of a State decision should be documented in a letter to the DOE, from the governor of the State, or another State official to whom the authority for this decision has been appropriately delegated. A copy of this letter must be sent to NRC.

The NRC has exclusive jurisdiction over both the radiological and non-radiological hazards of 11e.(2) byproduct material (NRC, 2000).

E3.0 THE LICENSE TERMINATION PROCESS

A licensee considering termination of its source material license should have in place an acceptable (by NRC) site decommissioning and reclamation plan and, if necessary, an acceptable ground-water corrective action program. This section describes the termination process that follows an NRC licensee completion of decommissioning, reclamation, and ground-water corrective action in accordance with the approved plans.

E3.1 Licensee Documentation of Completed Remedial and Decommissioning Actions

E3.1.1 Documentation of Completed Surface Remedial Actions

To ensure a timely and efficient NRC review, when reclamation of the tailings disposal cell is completed, the licensee should submit to NRC, for review, a report detailing the conduct and completion of the reclamation construction activities. This Construction Completion Report would consist primarily of a summary of quality assurance/quality control records and as-built drawings. A licensee may refer to the reports prepared by the DOE to document completion of remedial actions at Title I Project sites as guidance in developing its Construction Completion Report. However, some of the information presented in DOE reports (e.g., original design calculations) has been meant to ease the staff review rather than to meet documentation requirements.

If a Construction Completion Report or similar report is not submitted, it will be necessary for the staff to conduct a detailed technical review to meet its responsibilities under Section 83c of the Act. This review could involve several site visits and significant confirmation testing and would likely involve staff in the following technical disciplines: geotechnical engineering, surface water and erosion protection, radon barrier design, and soil radiation cleanup. Accurate

E–3

quality assurance/quality control records and photographs kept by a licensee during cell construction will be important input into the staff determination that reclamation has been conducted and completed in accordance with the approved plan.

If the NRC determines, as part of its review of the Construction Completion Report or during a site inspection, that a licensee has neglected to compile quality assurance/quality control records or has inadequate records, it may require the licensee to conduct appropriate sampling of those portions of the completed cell that are in question (e.g., of the radon barrier). If a licensee is unwilling or unable to comply, the staff or NRC contractors will conduct the sampling, and the costs involved will be included in the licensing and inspection fees assessed under 10 CFR 170.31. In addition, if a requirement to maintain quality assurance/quality control records is part of an approved reclamation plan, a licensee's lack of such records may be interpreted as a violation of the relevant license condition. This situation will be evaluated as part of the NRC inspection program. Appropriate NRC action would be taken in such instances.

E3.1.2 Documentation of Completed Site Decommissioning

Licensees are also required, under 10 CFR 40.42(j), to document the results of site decommissioning, which is done by conducting a radiation survey of the premises where the licensed activities were carried out. The results of this survey, the contents of which are specified at 10 CFR 40.42(j)(2), are submitted to NRC for review. A licensee has the option of demonstrating that the premises are suitable for release in a manner other than that specified at 10CFR 40.42. Additional documentation pertinent to site decommissioning and soil cleanup may be required by a specific license condition.

E3.1.3 Documentation of Completed Ground-Water Corrective Actions

Criteria 5A–5D, along with Criterion 13, of Appendix A incorporate the basic ground-water protection standards imposed by the U.S. Environmental Protection Agency (EPA) in 40 CFR Part 192, Subparts D and E (48 FR 45926, October 7, 1983). These standards apply during operations and before the end of closure. At a licensed site, if these ground-water protection standards are exceeded, the licensee is required to put into operation a ground-water corrective action program (Criterion 5D of Appendix A). The objective of the corrective action program is to return the hazardous constituent concentration levels to the concentration limits set as standards.

For licensees with continuing ground-water cleanup, NRC approval is required for the termination of corrective action. A licensee should submit appropriate ground-water monitoring data and other information that produce reasonable assurance that the ground water has been cleaned to meet the appropriate standards. This may include an application for alternate concentration limits if the licensee concludes that some alternate concentration limits for certain constituents are necessary. The staff will review alternate concentration limits in accordance with the most current version of the NRC staff technical position, "Alternate Concentration Limits for Title II Uranium Mills: Standard Format and Content Guide, and Standard Review Plan for Alternate Concentration Limit Applications" (NRC, 1996).

E3.2 NRC Review of Completed Closure Actions

On receipt of the Construction Completion Report, decommissioning report, ground-water completion report, or alternate concentration limit monitoring data, the staff will review the document for completeness and level of detail. Given a favorable finding, the staff will then review the content of the report for documentation that the action has been conducted in accordance with the license requirements and regulations. If that is the case, NRC will notify the licensee by formal correspondence, and, if the licensee so requests, amend the specific license, by deleting applicable license requirements for reclamation, decommissioning, or ground-water cleanup, and identifying requirements for any disposal cell observational period and/or environmental monitoring. The staff may conduct site inspections, examining first-hand the closure actions taken, including the quality assurance/quality control records.

Additionally, the staff should conduct a final construction completion inspection, which is expected to consist of a site walk-over and an examination of construction records. No independent verification of completed actions (e.g., confirmatory coring of the radon barrier) is expected, except on a case-by-case basis, as discussed previously.

With respect to construction of the tailings cell, the staff review of the Construction Completion Report, coupled with site inspections, will ensure that disposal cells are constructed in accordance with the approved design and plan (e.g., a summary of quality assurance/quality control records shows the appropriate number of material lifts have been placed).

The staff will rely on site inspections as the primary means of determining acceptable implementation of the licensee's approved decommissioning plan, especially in regard to soil cleanup. These inspections will consist of (1) reviews of procedures, (2) evaluations of procedure implementation, (3) evaluations of records and quality assurance, and (4) limited gamma surveys and soil sampling. In this way, the staff will gain the needed level of confidence in the licensee's performance to support its evaluation of the final decommissioning survey report. Confirmatory sampling, either by NRC or its contractors, should be conducted at sites for which additional confirmation beyond inspections is necessary. Specific criteria will be employed to identify those sites requiring confirmatory sampling.

E3.3 Observation Periods

E3.3.1 Following Completion of Surface Remedial Actions

The length of an observation period following completion of surface remediation will be determined on a site-specific basis, with a minimum period of 1 year, commencing at the completion of the erosion cover. Licensees should report significant cell degradation (e.g., the development of settlement or erosional features) occurring during this period.

Sites employing a full self-sustaining vegetative cover (Criterion 4 of Appendix A) may have an observation period of at least 2 years, and possibly as long as 5 years, based on specific site conditions and the requirements of 10 CFR Part 40, Appendix A.

Appendix E

A *de facto* observation period may exist at most sites where cleanup of ground-water contamination continues following the completion of surface reclamation (i.e., construction of the tailings disposal cell).

E3.3.2 Following Ground-Water Remediation

The reviewer should examine (1) ground-water completion reports, (2) ground-water corrective action reports, or (3) alternate concentration limit applications to verify that ground-water quality corrective actions have produced a stable water quality and that ground-water monitoring and analysis have been done to confirm the concentration of these contaminants in the ground water and to verify that they meet applicable standards. This should be done at the end of the 1-year stability ground-water monitoring period.

Ground-water stability monitoring and confirmation of constituents of concern will be acceptable if:

(1)   A one-time measurement of all constituents of concern has been collected and analyzed from all point-of-compliance wells. A constituent of concern is one that is (a) either (i) currently identified in 10 CFR Part 40, Appendix A, Criterion 13; or (ii) is not listed in Criterion 13, but is placed in a license condition as part of the staff review of the Corrective Action Plan; and (b) has been identified in the tailings liquor. NRC has flexibility to add other constituents not identified in Criterion 13. However, in identifying this second set of constituents, the staff should ensure that any additions are made based on a sound technical and regulatory basis. New constituents should be added in a timely manner, either when the corrective action plan is accepted for review, or at some time during the lifetime of the corrective action program. New constituents will not be required at the time of the license termination monitoring submittal.

Some examples of sound technical bases for new constituents follow:

(a)   For the Homestake/Grants and United Nuclear Corporation/Churchrock sites, the NRC staff, the DOE, and the EPA will work together to develop an interagency policy on closure and postclosure issues that will comply with the statutory and regulatory missions and requirements of all three agencies. For the Cotter/Canon City and UMETCO/Uravan sites, the State of Colorado is the primary regulatory authority and the NRC has a more limited role. Once all applicable NRC requirements are met, the NRC will have no basis for denying a request to terminate any specific license. However, before the NRC terminates any license for a site that is on the National Priority List or that is subject to continuing regulation by the EPA, the NRC will inform the DOE of the pending action, and where possible, will provide additional time for the DOE to resolve site issues it may have with the EPA.

(b)   Trends in ground-water contamination show that after several years of decreasing contamination, the levels of contamination begin to rise again.

(c)   Surrogate parameters that cover a family of constituents show an increase in the concentration in ground water. Therefore, the staff may require licensees to monitor for all constituents found in that family.

(d)   Some constituents used in the milling process, but not listed in Criterion 13, such as ammonia and nitrate, must be addressed.

Constituents should not be added just because an individual state regulatory body is concerned about that constituent. Having a state identify a constituent as one of concern to the state is not necessarily a proper basis for NRC to include that constituent.

(2)   The results of the one-time measurement sampling should be compared with the pre-operational applicable standards as specified in Criterion 5(C) or the license. All hazardous constituents must be shown to meet the standards specified in Criterion 5(C) or the license. If this measurement is taken sometime before license termination (3 or more years), the reviewer should ensure that recontamination has not occurred. This may be done by taking additional measurements or making analytical calculations.

(3)   The stability monitoring data should be inspected for any trends in increasing ground-water concentrations for those constituents of concern in the ground water that were being cleaned up by the corrective action plan.

If the staff reviews result in acceptance of confirmation and stability monitoring, the staff may conclude that:

(1)   The licensee has monitored all previously identified constituents of concern at the points of compliance.

(2)   The post-corrective action plan stability monitoring shows that the constituents of concern that were remediated will remain below compliance or alternate concentration limit standards.

(3)   The one-time sampling for constituents of concern shows that constituents of concern are below, and will remain below, relevant standards in 10 CFR Part 40, Appendix A, Criteria 5B(5) and 5B(6).

(4)   All ground-water corrective action programs have ceased operation.

(5)   All identified constituents of concern for which compliance sampling is being conducted have been returned to the concentration limits set as standards.

E3.4   Long-Term Site Surveillance Funding

Before termination of the specific license, NRC will set the final amount of the long-term site surveillance charge to be paid by the licensee in accordance with Criterion 10 of

10 CFR Part 40, Appendix A. The NRC process for determining this amount will include consultations with the licensee and the custodial agency. This charge must be paid to the United States general treasury or to the appropriate state agency before the specific license can be terminated.

Notify NRC and custodial agency 2 years before estimated date of license termination. When a licensee has completed site reclamation, decommissioning, and, if necessary, ground-water corrective action, and is ready to terminate its specific source material license, it must formally notify NRC of its intentions. Such notification should be accompanied by a completed NRC Form 314, "Certificate of Disposition of Materials" or approved alternate.

E3.4.1 Bases for Determination of Surveillance Charge

The basic criterion for tailings disposal is to avoid dependance on perpetual human care and on-going maintenance to preserve the isolation of the tailings. NRC, in Criterion 1 of Appendix A, concludes that:

> The general goal or broad objective in siting and design decisions is permanent isolation of tailings and associated contaminants by minimizing disturbance and dispersion by natural forces, and to do so without ongoing maintenance.

However, as further indicated in Criterion 1, for practical purposes, specific design and siting considerations must involve finite time limits. For this reason, Criterion 6 contains longevity standards for design of disposal cells.

In order that the isolation of the tailings and associated contaminants can be preserved to the extent possible, the Atomic Energy Act, as amended, provides that title to the byproduct material and associated land be transferred to the care of the United States, the State, or the tribe, as discussed previously. NRC has interpreted such long-term custody by a governmental agency, whether Federal or State, as "a prudent, added measure of control" (NRC, 1980), so that land uses that might contribute to the degradation of the cover or lead to direct human exposures can be prevented.

In the "Final Generic Environmental Impact Statement on Uranium Milling" (NRC, 1980), NRC staff developed the bases for the long-term surveillance charge, given the intent that no ongoing active maintenance of site conditions should be necessary to preserve waste isolation. In the final generic environmental impact statement, the following actions are assumed for the "passive monitoring" approach to surveillance of the site are as follows:

(1)   An annual visual inspection of the site, either as a site visit or a visual inspection from an aircraft, should be conducted.

(2)   No maintenance of equipment or facilities, no fence replacement, no sampling, and no airborne environmental monitoring would be expected.

(3)   Essentially, the only costs for continued surveillance/maintenance would consist of time spent in preparing for the inspection, travel to the site, conduct of the inspection, and annual report writing.

(4)   Minimal NRC oversight would be required.

Passive monitoring, thus, would not involve such activities as irrigation, hauling of fill, regrading, or seeding.

Finally, as discussed previously, licensees will contribute the funds necessary to cover the costs of long-term surveillance of their sites. The charge assessed is a one-time fee, which will yield interest on the funds, assuming a 1-percent annual real interest rate, sufficient to cover the annual costs of site surveillance. The final generic environmental impact statement contains more detailed discussion regarding the determination of this interest rate.

E3.4.2 Determination of Surveillance Charge Amount

On the basis of the assumptions discussed in Section E3.4.1 (above), NRC developed the minimum long-term surveillance charge of $250,000 (1978 dollars) as stated in Criterion 10 of Appendix A. It is this charge, adjusted to account for inflation, that the licensee is required to pay into the general treasury of the United States, or alternately, to the appropriate State agency (if the State is to become the long-term site custodian). The methodology the staff will use to determine the adjusted surveillance charge that accounts for inflationary increases since 1978 includes (1) using the Consumer Price Index available at the time the licensee requests termination and (2) applying the rate of increase for the last month for which it has been calculated to any following month leading to license termination. For example, in June 1996, NRC determined the final surveillance charge for the TVA/Edgemont site. In doing this, the staff used the April 1996 Consumer Price Index and applied the rate of increase between March and April to the following months.

Criterion 10 allows for the escalation of this minimum charge if, on the basis of a site-specific evaluation, the expected site surveillance or control requirements are determined to be significantly greater than those specified in Criterion 12 of Appendix A (i.e., annual inspections to confirm site integrity and determine the need, if any, for maintenance or monitoring).

Escalation could result from a licensee's proposal of alternatives to the requirements in Appendix A, as allowed under Section 84c of the Atomic Energy Act, as amended. For example, a licensee could demonstrate by analysis that the only mechanism for achieving a minimum disposal cell design life of 200 years at its site is through the use of ongoing maintenance. NRC may approve such a design if it finds that the design will achieve a level of stabilization and containment for the site concerned, and a level of protection of public health and safety, and of the environment, that is equivalent to, to the extent practicable, or more stringent than, the level of protection that would be achieved by meeting NRC requirements. However, the licensee would likely be required to place additional funds in the long-term surveillance charge to cover the costs of the ongoing maintenance.

Appendix E

Another situation that may lead to the escalation of the minimum charge is the recognition that some degree of active care (e.g., vegetation control, maintenance of erosional control measures) is necessary to preserve the as-designed conditions of the site. This need should become apparent in the course of site observations during the reclamation and observational periods.

In any case, any escalation in the minimum charge will be discussed with the licensee and long-term custodian, before license termination. Any final variance in the funding requirements will be determined solely by NRC.

A situation may arise in which the custodial agency wants to have commitments in the long-term surveillance plan that are beyond those required in Appendix A and that NRC does not determine are necessary. In such a case, the amount of the long-term surveillance charge would not be affected (NRC, 1990, "Detailed Comment Analysis," Comment 1.2). The custodial agency must identify a mechanism for funding these additional self-imposed requirements.

E3.4.3 Payment of Long-Term Surveillance Charge

Licensees may pay the final site surveillance charge to the NRC or the custodial agency. If paid to NRC, the funds will be deposited, in accordance with the Miscellaneous Receipts Act, in the U.S. General Treasury. A custodial agency receiving payment from the licensee will need to document receipt and subsequent deposition of the payment. Copies of such documentation should be sent to NRC.

E3.5   Preparation of the Long-Term Surveillance Plan

While surface remediation and ground-water cleanup activities are ongoing, it is in the best interest of the licensee to contact the custodial agency with regard to that agency preparation of the site long-term surveillance plan. The custodial agency responsibilities under the general license are defined in the long-term surveillance plan, the required contents of which are provided at 10 CFR 40.28 and in Criterion 12 of Appendix A, as follows:

- A legal description of the site to be transferred and licensed

- A detailed description of the site, as a baseline from which future inspectors can determine the nature and seriousness of any changes {licensees may reference previously submitted information, to the extent applicable, in providing this description [10 CFR 40.31(a)]}

- A detailed description of the long-term surveillance program, including (1) the frequency of inspections and reporting to the NRC; (2) the frequency and extent of ground-water monitoring, if required; (3) appropriate ground-water concentration limits; and (4) inspection procedures and personnel qualifications

- The criteria for follow-up inspections in response to observations from routine inspections or extreme natural events

- The criteria for instituting maintenance or emergency measures

In addition to these regulatory requirements (also see Appendix D of this standard review plan), NRC will also require that the long-term surveillance plan contain documentation of title transfer of the site from the licensee to the custodial agency. This requirement does not apply to sites located on tribal lands, since transfer does not occur for such sites (Criterion 11F of Appendix A).

Because the long-term surveillance plan must reflect the remediated condition of the site, it is expected that the existing licensee will work with the custodial agency to prepare the long-term surveillance plan. As discussed in Section E2.2 (above), this will likely involve supplying the custodial agency with appropriate documentation (e.g., as-built drawings) of the remedial actions taken and reaching agreements (formal or informal) with the custodial agency regarding the necessary surveillance control features of the site (e.g., boundary markers, fencing).

As the likely custodial agency for most, if not all, of the sites, the DOE has developed a generic long-term surveillance plan shell. For sites under the long-term care of the DOE, significant portions of the long-term surveillance plan will not change from site to site (e.g., criteria for follow-up inspections and for instituting maintenance or emergency measures). Therefore, the staff may focus its review on the site-specific information in the long-term surveillance plan. This information would include site-specific activities that are not to be reflected in the long-term care charge, but are voluntarily committed to by the custodial agency.

E3.6   Termination of the Specific License/Issuance of the General License

Actual termination of a licensee-specific license and the subsequent placement of the site under the general license provisions of 10 CFR 40.28 will involve a number of separate actions to be completed by the NRC. Significant internal coordination (and external, if Agreement State licensees are involved) will be required so that these actions will be completed in an efficient and timely manner, thereby ensuring that the byproduct material and any land used for the disposal of such byproduct material remain under NRC license throughout the process.

E3.6.1 NRC Determination Under Section 83c of the Act

Under Section 83c of the Atomic Energy Act, as amended, NRC must determine whether all applicable standards and requirements have been met by the licensee in the completion of site reclamation, decommissioning, and ground-water corrective action before a licensee's license can be terminated. Necessarily, this determination will rely primarily on NRC reviews and acceptance of the documentation presented by the licensee. In addition, NRC site closure inspection activities, potentially including limited confirmatory radiological surveys, will provide supplemental information for NRC determination.

E3.6.2 NRC Review and Acceptance of the Long-Term Surveillance Plan

A long-term surveillance plan is required before termination of the specific license and placement of the site and byproduct material under the 10 CFR 40.28 general license. Review

Appendix E

and acceptance of the long-term surveillance plan is the sole purview of NRC. Lack of NRC acceptance of a site long-term surveillance plan can delay termination of the specific license.

NRC staff acceptance of a long-term surveillance plan will be documented in written notification to the custodial agency, and, separately, by noticing the action in the *Federal Register*.

E3.6.3 Issuance of a Specific Order Under 10 CFR 40.28

If NRC has not received an acceptable long-term surveillance plan for a reclaimed site ready for transfer to the custodial agency, the agency has two options available to it. First, if appropriate, the Commission may choose not to terminate the existing license for a short period of time, while awaiting an acceptable long-term surveillance plan. Alternately, under 10 CFR 40.28, NRC may issue a specific order to the custodial agency to take custody of the site and to commence long-term surveillance, while the agency prepares the long-term surveillance plan for final NRC approval.

NRC would require substantial basis to support issuance of an order. The basis would include an understanding of the circumstances leading to the custodial agency inability to take the site. Factors that the NRC would consider include whether:

(1)   Adequate notice (at least 16 months) has been provided by the existing licensee to allow the custodial agency to effect title transfer to the land and byproduct material.

(2)   Sufficient time (at least 2 years) has been allowed for the custodial agency to prepare, and the NRC to review, the long-term surveillance plan.

(3)   NRC has reviewed the Construction Completion Report, decommissioning report, and ground-water cleanup report and has conducted the final license-termination inspection and found that the closure actions were completed in an acceptable manner.

(4)   Site degradation has occurred, and whether appropriate repairs have been completed.

(5)   The required long-term surveillance funding payments have been made to the U.S. General Treasury or to the designated state agency.

(6)   The custodial agency has an acceptable rationale for delaying inclusion of the site under the general license.

In cases in which the DOE or another presidentially designated Federal agency will be the long-term custodian and is unable to take custody of the site because of lack of funding, NRC may still order the agency to take custody. The intended custodial agency will have at most 1 year (i.e., the time by which an annual site inspection is to have been completed) in which to obtain the funds through the necessary appropriations process.

E3.6.4 Transfer of Site Control to the Custodial Agency

Given a determination that all applicable standards and requirements have been met and acceptance of the site long-term surveillance plan, NRC will need to complete the following remaining relevant licensing actions: (1) terminating the specific license by letter of termination addressed to the specific licensee, or concurring in the Agreement State termination of the specific license; (2) placing the site under the general license in 10 CFR 40.28; (3) noticing, in the *Federal Register,* the completion of these licensing actions; and (4) informing appropriate Federal and State officials directly of the termination of the specific license and the placement of the site under the general license.

The long-term custodian, for its part, should be prepared to accept title to the land and byproduct material. These final actions should be completed within a relatively short period of time (i.e., within a week).

E4.0 ADDITIONAL ISSUES

E4.1 Uranium Mill Tailings Radiation Control Act of 1978, As Amended, Title II Sites on Tribal Lands

For Uranium Mill Tailings Radiation Control Act of 1978,, as amended, Title II disposal sites on tribal lands, long-term surveillance will be accomplished by the Federal government. The custodial agency is required to enter into arrangements with NRC to ensure this surveillance. The Uranium Mill Tailings Radiation Control Act of 1978, as amended, does not state explicitly which Federal agency is responsible for the disposal site. In addition, because these sites are located on tribal lands, no title transfer will occur.

Currently, the only site on tribal lands is the previous Western Nuclear, Inc., Sherwood uranium mill, located on the Spokane Indian tribe reservation in eastern Washington State. The Western Nuclear, Inc. Sherwood license was terminated in early 2001. Under long-term care, arrangements for the Sherwood site involve a site access agreement between the Indian tribe and DOE. DOE is allowed to conduct the required site surveillance and the site is accessible to NRC.

E4.2 Exclusive Jurisdiction

The Commission has determined that NRC has exclusive jurisdiction over both the radiological and non-radiological hazards of 11e.(2) byproduct material (NRC 2002). Notwithstanding this determination, the NRC staff intends to work with the states on issues related to a licensee's completion of remedial actions and preparation for license termination. Although the NRC will, to the extent possible, accommodate a state's interest or perspective, it retains the right to terminate a specific license should a licensee have completed closure activities in accordance with NRC-approved closure plans.Where the issues involved are not those of direct NRC concern, NRC will address such issues with the states or other federal agencies on a case-by-case basis.

Currently, four sites (two NRC licensees: the United Nuclear Corporation/Church Rock site and the Homestake Mining Company/Grants site; and two Agreement State licensees: the Cotter Corp./Canon City and the UMETCO/Uravan sites, both in Colorado) are on the Superfund National Priorities List. For these sites, NRC will work with states and Superfund administering agencies to determine if it is appropriate to terminate the licenses.

E5.0 REFERENCES

NRC. "Termination of Uranium Milling Licenses in Agreement States." STP SA–900 Procedure. Washington, DC: NRC, Office of State and Tribal Programs.   December 2002.

———. SECY–99–0277. "Concurrent Jurisdiction of Non-Radiological Hazards of Uranium Mill Tailings." Washington, DC: NRC. August 2000.

———. "Staff Technical Position: Alternate Concentration Limits for Title II Uranium Mills: Standard Format and Content Guide, and Standard Review Plan for Alternate Concentration Limit Applications." Washington, DC: NRC. February 1996.

———. SECY–90–282. "Rulemaking Issue (Affirmation): Amendments to 10 CFR Part 40 for General Licenses for the Custody and Long-Term Care of Uranium and Thorium Mill Tailings Disposal Sites." Washington, DC: NRC. August 1990.

———. NUREG–0706, "Final Generic Environmental Impact Statement on Uranium Milling." Washington, DC: NRC. 1980.

APPENDIX F

GUIDANCE TO THE U.S. NUCLEAR REGULATORY COMMISSION STAFF ON EFFLUENT DISPOSAL AT LICENSED URANIUM RECOVERY FACILITIES: CONVENTIONAL MILLS

F1.0 BACKGROUND

U.S. Nuclear Regulatory Commission (NRC)-licensed uranium mill recovery facilities produce liquid wastes (i.e., effluent) that require proper disposal. NRC Office of Nuclear Material Safety and Safeguards policy is presented below.

F1.1 Purpose and Applicability

This appendix presents guidance and discusses the technical and regulatory basis for review and evaluation of applications for the disposal of liquid waste. It is primarily intended to guide NRC staff reviews of site-specific applications for disposal of liquid waste.

F1.2 On-Site Evaporation

Applications for on-site evaporation systems must demonstrate that the proposed disposal facility is designed, operated, and closed in a manner that prevents migration of waste from the evaporation systems to subsurface soil, ground water, or surface water in accordance with 10 CFR Part 40, Appendix A. Applicants must also demonstrate that site-specific ground-water protection standards and monitoring requirements are adequately established to detect any migration of contaminants to the ground water and to implement corrective action to restore ground-water quality if, and when, necessary, as required by the regulations.

If surface impoundments are employed for liquid waste and solid wastes (sludge) that are 11e.(2) byproduct material. Licensees must demonstrate that surface impoundments are designed, operated, and decommissioned in a manner that prevents migration of waste from the surface impoundment to subsurface soil, ground-water, or surface water in accordance with 10 CFR Part 40, Appendix A. Applicants must also demonstrate that monitoring requirements are adequately established to detect any migration of contaminants to the ground-water. Solid waste material must be disposed of in an existing tailings impoundment or 11e.(2) disposal cell in accordance with 10 CFR Part 40, Appendix A, Criterion 2.

If surface impoundments are employed for evaporation, but they are not used for waste disposal, they must comply with the design provisions for surface impoundments [Criterion 5A(1) through Criterion 5A(5)]; measures for ground-water protection programs (Criterion 5E); and seepage control (Criterion 5F) of 10 CFR Part 40, Appendix A. However, if surface impoundments are employed for evaporation and waste disposal, they must comply with the regulatory requirements in 10 CFR Part 40, Appendix A. These include the design provisions for surface impoundments [Criterion 5A(1) through Criterion 5A(5)]; measures for ground-water protection programs (Criterion 5E); and seepage control (Criterion 5F). In addition, evaporation ponds must also meet other generally applicable regulatory provisions in Appendix A, in particular, the site-specific ground-water protection standards and leak detection requirements (Criterion 5B and Criterion 5C); corrective action programs (Criterion 5D); ground-water monitoring requirements (Criterion 7); and closure requirements (Criterion 6).

Appendix F

F1.3 Release in Surface Waters

Process waste water resulting from operations is 11e.(2) byproduct material. The U.S. Environmental Protection Agency (EPA), in accordance with 40 CFR 440.34, does not allow new facilities to discharge process waste water to navigable waters. For release of this waste to surface waters, existing licensees must meet the requirements of 10 CFR 20.1302(b)(2), and should demonstrate that doses are maintained as low as reasonably achievable (ALARA). NRC has no specific requirements for non-radiological constituents, and may adopt the appropriate State limits. Anticipated discharge must be described in enough detail to evaluate environmental impacts. Appropriate State and Federal agency permits should be obtained in accordance with 10 CFR 20.2007.

F1.4 Land Applications

For the land application of process waste water, the applicant must meet the regulatory provisions in 10 CFR 20.2002 and demonstrate that doses are ALARA and within the dose limits in 10 CFR 20.1301. Proposed land application activities should be described in sufficient detail to satisfy the NRC need to assess environmental impacts. This may require analysis to assess the chemical toxicity of radioactive and nonradioactive constituents. Specifically, licensees must provide: (i) a description of the waste, including its physical and chemical properties that are important to risk evaluation; (ii) the proposed manner and conditions of waste disposal; (iii) projected concentrations of radioactive contaminants in the soil; and (iv) projected impacts on ground-water and surface-water quality and on land uses, especially crops and vegetation. In addition, projected exposures and health risks that may be associated with radioactive constituents reaching the food chain must be analyzed to ensure that doses are ALARA. Proposals should include provisions for periodic soil surveys to verify that contaminant levels in the soil do not exceed those projected, and should also include a remediation plan that can be implemented if projected levels are exceeded. Appropriate State and Federal agency permits must be obtained in accordance with 10 CFR 20.2007. The applicant must also comply with NRC regulatory provisions for decommissioning. The applicant should also address whether the proposed land applications methodologies will comply with 10 CFR Part 40, Appendix A, Criterion 6(6), at the time of decommissioning.

F1.5 Deep Well Injection

Proposals for disposal of liquid waste from process water by injection in deep wells must meet the regulatory provisions in 10 CFR 20.2002 and demonstrate that doses are ALARA and within the dose limits in 10 CFR 20.1301. The injection facility should be described in sufficient detail to satisfy the NRC need to assess environmental impacts. Specifically, proposals must include: (i) a description of the waste, including its physical and chemical properties important to risk evaluation; (ii) the proposed manner and conditions of waste disposal; (iii) an analysis and evaluation of pertinent information on the nature of the environment; (iv) information on the nature and location of other potentially affected facilities; and (v) analyses and procedures to ensure that doses are ALARA and within the dose limits in 10 CFR 20.1301.

In addition, pursuant to the provisions of 10 CFR 20.2007, proposals for disposal by injection in deep wells should also meet any other applicable Federal, State, and local government regulations pertaining to deep well injection. Applicants must obtain any necessary permits for this purpose. In particular, proposals must satisfy the EPA regulatory provisions in 40 CFR Part 146: Underground Injection Control (UIC) Program: Criteria and Standards, and applicants must obtain necessary permits from EPA and/or States authorized by EPA to enforce these provisions. In general, applications that satisfy EPA regulations under the UIC Program, which are approved by the EPA or an EPA-authorized State issuing the UIC permit and the applicable provisions of 10 CFR Part 20, will also be approved by the staff. Licensees and applicants disposing of liquid waste from process water by injection in deep wells are further required to comply with NRC regulatory provisions for decommissioning.

APPENDIX G

NATIONAL HISTORIC PRESERVATION ACT AND ENDANGERED SPECIES ACT CONSULTATIONS

G1.0 BACKGROUND

The National Historic Preservation Act requires federal agencies to consider the effects of actions licensed by Federal agencies on properties included in or eligible for the *National Register of Historic Places*. The reclamation of a mill could impact historic properties directly (e.g., destruction or alteration of the integrity of a property) or indirectly (e.g., prohibiting access or increasing the potential for vandalism). Similarly, the Endangered Species Act requires that federal agencies consult with the U.S. Fish and Wildlife Service on any Federal action that could impact endangered species or their habitats. This appendix presents guidance to the U.S. Nuclear Regulatory Commission (NRC) staff on how to fulfill the NRC obligations under the National Historic Preservation and Endangered Species Acts.

G2.0 NATIONAL HISTORIC PRESERVATION ACT

G2.1 Review Procedures

The reviewer should ensure that those historic and cultural resources that could be impacted by proposed mill reclamation have been identified, located, and described in sufficient detail to serve as the basis for subsequent analysis and assessment of these impacts. Historic and cultural resources include districts, sites, buildings, structures, or objects of historical, archaeological, architectural, or cultural significance. The staff should review the results of any surveys conducted by the applicant, the location and significance of any properties that are listed in or eligible for inclusion in the *National Register of Historic Places (National Register)* as a historic place, and any additional information pertaining to the identification and description of historic properties that could be impacted by reclamation of the proposed mill. The descriptions to be examined by this review should be of sufficient detail to permit staff assessment and evaluation of specific impacts to historic and cultural resources from reclamation of the mill.

Regulatory criteria for the review of the historic properties that could be impacted by proposed reclamation of mills are based on the relevant requirements of the following:

- 36 CFR Part 800 defines the process by which a Federal agency conforms to the requirements under Sections 106 and 110 of the National Historic Preservation Act to ensure that agency-licensed undertakings consider the effects of the undertaking on historic properties included in or eligible for the *National Register*. Under this regulation, the Federal agency is required to identify and evaluate all historic properties in the project areas and take measures to mitigate adverse affects.

- 36 CFR Part 63 contains guidance by which historic properties are evaluated and determined eligible for listing on the *National Register*.

Appendix G

The reviewer should take the following steps to obtain the necessary information:

(1)   Contact the appropriate State historic preservation officer to determine if there are any historic or cultural properties near the proposed mill site. In areas of Indian tribal land, the Indian tribal agencies may act as the State historic preservation officer. State historic preservation officer and tribal historic preservation officer lists are found on the Advisory Council on Historic Preservation Internet Home Page (http://www.achp.gov).

    (a)   NRC can authorize the applicant to initiate consultation with the State historic preservation officer or tribal historic preservation officer, but remains legally responsible for all findings. Notify the State historic preservation officer/tribal historic preservation officer when an applicant is so authorized.

    (b)   Make initial contact by phone and invite the State historic preservation officer to participate in the site visit. Then request information from the State historic preservation officer by letter.

    (c)   If the State historic preservation officer has comments or information that add to or amplify information given to the applicant, request that the State historic preservation officer forward, by letter to the staff, these additional comments.

(2)   Contact the Archeology and Ethnography Program of the National Park Service, U.S. Department of Interior (http://www.cr.nps.gov/aad/). This office has expertise in the area of historic and cultural preservation and is staffed with professionals who can assist in the environmental review and in analyzing the results of applicant surveys and investigations.

(3)   In consultation with the State historic preservation officer, apply the *National Register* criteria outlined by the U.S. Department of the Interior (National Park Service, 1990, 1991) to all identified historic properties that are on the facility site or that will be directly affected by facility construction. If a property appears to meet the criteria, or if it is questionable whether the criteria are met, the staff should request, in writing, an opinion from the U.S. Department of the Interior about the property eligibility for inclusion in the *National Register*. The request for determination of eligibility should be sent directly to the Keeper of the National Register of Historic Places, National Park Service, U.S. Department of the Interior, Washington, DC, 20013-7127.

(4)   Have the National Park Service–Archeology and Ethnography Program staff assist in defining the requirements of additional surveys and investigations that the staff decides should be completed by the applicant and in reviewing the results of these surveys.

(5)   Consult the *National Register* to verify the list of *National Register* properties presented by the licensee. Since a proposed facility can have a visual or audible effect on historic and cultural resources that are located some distance from the proposed facility site, all *National Register* properties within the area of potential effects of the proposed facility or off-site areas should be identified.

(6)   Discuss with the State historic preservation officer and, where appropriate, the State archaeologist and State historian, the information presented to the aplicant by the State historic preservation officer. The State historic preservation officer can alert the staff to relevant State and local laws, orders, ordinances, or regulations aimed at the preservation of cultural resources within the licensee State. Discuss with the State historic preservation officer any organizations or individuals that might be able to assist in identifying and locating archaeological and historic resources (e.g., university, Indian tribal archaeological and historical staffs, and regional U.S. Bureau of Land Management archaeologists).

(7)   To discourage property vandalism and scavenging, it is necessary to present information on location of artifacts to the State historic preservation officer in a confidential manner. Summary information, which does not include site-specific information, could be included in the licensee and NRC staff documentation.

(8)   Contact the Advisory Council on Historic Preservation if guidance is needed, if there are substantial impacts on important properties, in the event of a disagreement, or if there are issues of concern to Indian tribes.

For license renewals and amendment applications, Appendix A to this standard review plan provides guidance for examining facility operations and the approach that should be used in evaluating amendments and renewal applications.

G2.2 Acceptance Criteria

The kinds of data and information needed will be affected by site- and facility-specific factors and the degree of detail should be modified according to the anticipated magnitude of the potential impact. Guidance can be found on the National Park Service Internet Home Page at http://www.cr.nps.gov/nr/nrpubs.html and in NUREG–1748 (NRC, 2001). The licensee should present the following data or information:

(1)   A detailed description of any archaeological or historical surveys of the proposed site, including the following:

    (a)   The physical extent of the survey: If the entire site was not surveyed, the basis for selecting the area to be surveyed.

    (b)   A brief description of the survey techniques used and the reason for the selection of the survey techniques used.

    (c)   The qualifications of the surveyors.

    (d)   The findings of the survey in sufficient detail to permit a subsequent independent assessment of the impact of the proposed project on archaeological and historic resources.

(2)   The results of consultation with Federal, State, local, and affected Indian tribal agencies.

(3)   The comments of any organizations contacted by the licensee to locate and assess archaeological and historic resources located on or near the proposed mill site.

(4)   A description of any historic property within the area of potential effects of the mill that are in or have been determined eligible for inclusion in the *National Register* or that are included in state or local registers or inventories of historic and cultural resources.

(5)   A map indicating the location of all identified historic landmarks and historic places with respect to the location of facilities such as buildings, new roads, well fields, pipelines, surface impoundments, and utilities (not to be made public).

(6)   A license condition prohibiting work if cultural artifacts are found in locations other than those already identified.

(7)   The likely impact of the presence of new roads, pipelines, or other utilities on historic and cultural resources.

(8)   A rating of the aesthetic and scenic quality of the site in accordance with the U.S. Bureau of Land Management Visual Resource Management System (U.S. Bureau of Land Management, 1984, 1986a,b).

(9)   The following information should usually be briefly described in the environmental assessment:

    (a)   Historic properties listed in or eligible for inclusion in the *National Register*. Any resource considered to be eligible for the *National Register* should have concurrence from the appropriate state historic preservation officer.

    (b)   Historic properties included in state or local registers or inventories.

    (c)   Any additional important historic or cultural properties.

    (d)   The efforts to locate and identify previously recorded archaeological and historic sites.

    (e)   The overall results and adequacy of any surveys (archival or field) that were conducted by the applicant.

    (f)   A list of organizations and individuals contacted by the applicant or the NRC staff who provided significant information concerning the location of historic and cultural properties.

G2.3 Evaluation Findings

If the staff review results in the acceptance of the characterization of the historic and cultural resources, the following conclusions may be presented in the environmental assessment.

The staff has completed its review of the site characterization information concerned with historic and cultural resources near the _____ uranium mill facility. This review included an evaluation using the review procedures and the acceptance criteria outlined in Sections 2.1 and 2.2 of Appendix G of NUREG–1620. The licensee has acceptably described the historic and cultural resources near the site. A listing of all nearby areas and properties included or eligible for inclusion in the *National Register of Historic Places* is provided. A map indicates where all historic and cultural resources are located with respect to facilities. A record of the investigation of places and properties with historic and cultural significance, which follows guidance equivalent to that of the National Park Service, is provided. Contact with local tribal authorities is acceptably documented. A letter from the State historic preservation officer addressing any issues related to the properties that might be affected by the reclamation is included. The licensee has acceptably demonstrated that the State historic preservation officer and tribal authorities agree with the planned protection from, or determination of lack of conflict with, facilities and activities and with any places of importance to the State, Federal, or tribal authorities. The licensee has acceptably rated the aesthetic and scenic quality of the site in accordance with the U.S. Bureau of Land Management Visual Resource Inventory and Evaluation System.

On the basis of the information presented in the application, and the detailed review conducted of the characterization of historic and cultural resources near the _____ uranium mill facility, the staff concludes that the information is acceptable and is in compliance with 10 CFR 51.45, which requires a description of the affected environment containing sufficient data to aid the Commission in its conduct of an independent analysis.

G2.4 References

10 CFR 51.45, "Environmental report."

36 CFR Part 63, "Determination of Eligibility for Inclusion in the National Register of Historic Places."

36 CFR Part 800, "Protection of Historic and Cultural Properties."

National Historic Preservation Act, as amended. 16 USC 470 *et seq.*

National Park Service. "How to Apply the National Register Criteria for Evaluation." *National Register.* Bulletin No. 15. Washington, DC: U.S. Department of the Interior. 1991.

———. "Guidelines for Evaluating and Documenting Traditional Cultural Properties." *National Register.* Bulletin No. 38. Washington, DC: U.S. Department of the Interior. 1990.

——. "Guidelines for Local Surveys: A Basis for Preservation Planning." *National Register*. Bulletin No. 24. Washington, DC: U.S. Department of the Interior. 1985.

NRC. NUREG–1748, "Environmental Review Guidance for Licensing Actions Associated with NMSS Programs." Draft. Washington, DC: NRC. 2001.

U.S. Bureau of Land Management. "Visual Resource Inventory." BLM Report H–8410–1. 1986a.

——. "Visual Resource Contrast Rating." BLM Report H–8431–1. 1986b.

——. "Visual Resource Management." BLM Report 8500. 1984.

U.S. Department of the Interior. "Archeology and Historic Preservation; Secretary of the Interior's Standards and Guidelines." 48 FR 44716. pp. 44,716–44,742. 1983.

G3.0 ENDANGERED SPECIES ACT

G3.1 Review Procedures

Federal agencies must determine if any proposed actions could impact endangered species or their habitats. For uranium mills, the NRC staff should take into consideration impacts resulting from excavation of clay used in constructing radon barriers or procurement of rocks used in riprap. Other surface reclamation work, such as the cleanup of windblown tailings, has the potential to impact endangered animals or plants. Also, the staff should review the processing of any alternate concentration limit application if the proposed site is located on or near surface water that contains endangered animal or plant species.

Procedures for conducting consultations with the U.S. Fish and Wildlife Service are contained in the endangered species consultation handbook (U.S. Fish and Wildlife Service/National Marine Fisheries Service, 1998). The reviewer analysis should consist of the following steps:

(1)   Contact the U.S. Fish and Wildlife Service regional office or field office to obtain the list of threatened or endangered plant and animal species that may be present near the site. The attached table indicates the States in each U.S. Fish and Wildlife Service regional office and provides contact information.

(2)   The licensee may request the species list; however, the NRC must formally designate the licensee in writing (U.S. Fish and Wildlife Service/National Marine Fisheries Service, 1998, pp. 2–13).

(3)   If there may be endangered or threatened animal or plant species on or near the site, the reviewer should discuss the proposed action with the U.S. Fish and Wildlife Service and may need to ask the licensee to perform a survey and a biological assessment

(50 CFR 402.12) to evaluate the potential effects of the action on threatened and endangered species. Either the NRC or the licensee can prepare the biological assessment (U.S. Fish and Wildlife Service/National Marine Fisheries Service, 1998, pp. 3–11).

(4)   Each State should be consulted about its own procedures for considering impacts to state-listed species.

G3.2 Acceptance Criteria

Consultations on identifying threatened and endangered species under the Endangered Species Act of 1973, as amended, are acceptable if the following criteria are met:

(1)   The environmental impact assessment provides sufficient information to ensure that the licensing action is not likely to jeopardize the continued existence of any endangered species or threatened species or result in the destruction or adverse modification of the habitat of such species. The demonstration of compliance with this objective requires consultation with the U.S. Fish and Wildlife Service and/or the National Marine Fisheries Service. The U.S. Fish and Wildlife Service will coordinate with the National Marine Fisheries Service.

(2)   The licensee provides adequate discussion of the status of compliance with applicable permits, licenses and other environmental requirements that have been imposed by Federal agencies.

(3)   There is adequate information on interagency cooperation and consultations with Federal, State, and local agencies with regard to the Endangered Species Act.

G3.3 Evaluation Findings

If staff review results in acceptance of compliance with the Endangered Species Act, the following conclusions may be presented in the environmental assessment:

The staff has completed its review of the site characterization information concerned with the threatened and endangered species near the _____ uranium mill facility. This review included an evaluation using review procedures and acceptance criteria outlined in Sections 3.1 and 3.2 of Appendix G of NUREG–1620. The licensee has acceptably described the presence of threatened and endangered species near the site. Consultations with the U.S. Fish and Wildlife Service and/or the National Marine Fisheries Service on threatened and endangered species were conducted and are documented in an acceptable manner. Any impacts on these species and their habitats have been identified, and mitigation measures necessary to avoid adverse impacts have been described.

On the basis of the information presented in the application, and the detailed review conducted of the characterization of threatened and endangered species near the _____ uranium mill facility, the staff concludes that the information is acceptable and is in compliance with 10 CFR 51.45, which requires a description of the affected environment containing sufficient data to aid the Commission in its conduct of an independent analysis.

G3.4 References

10 CFR 51.45, "Environmental Report."

40 CFR 1502.25, "Environmental Review and Consultation Requirements."

50 CFR Part 402, "Interagency Cooperation-Endangered Species Act of 1973, as Amended."

Endangered Species Act of 1973, as amended, 16 USC 1531 *et seq.*

U.S. Fish and Wildlife Service/National Marine Fisheries Service. "Endangered Species Consultation Handbook, Procedures for Conducting Consultation and Conference Activities Under Section 7 of the Endangered Species Act." Washington, DC: U.S. Fish and Wildlife Service/National Marine Fisheries Service. 1998.

U.S. Fish and Wildlife Service
Endangered Species Program Contacts

Washington, DC Office

U.S. Fish and Wildlife Service
Division of Endangered Species
Mail Stop 420ARLSQ
1849 C St., N.W., Washington, DC 20240
http://www.fws.gov

Region One (CA, HI, ID, NV, OR, WA)

Chief, Division of Endangered Species
Eastside Federal Complex, 911 NE 11th Ave, Portland, OR 97232
http://pacific.fws.gov
Contact: Call field office

Region Two (AZ, NM, OK, TX)

Chief, Division of Endangered Species
P.O. Box 1306, Albuquerque, NM 87103
http://southwest.fws.gov
Contact:: Species list on Internet for each county
Call field supervisor if there are questions

Region Three (IA, IL, IN, MI, MN, MO, OH, WI)

Chief, Ecological Services Operations
Federal Building, Ft. Snelling, Twin Cities, MN 55111
http://midwest.fws.gov
Contact: Species list on Internet by county, (except Missouri)
Call field supervisor if there are questions or if the site is in Missouri

Region Four (AL, AR, FL, GA, KY, LA, MS, NC, PR, SC, TN, U.S. VI)

Programmatic Assistant
Regional Director for Ecological Services
1875 Century Blvd., Suite 200, Atlanta, GA 30345
http://southeast.fws.gov
Contact: Letter to Programmatic Assistant

Region Five (CT, DC, DE, MA, MD, ME, NH, NJ, NY, PA, RI, VA, VT, WV)

Chief, Division of Endangered Species
300 Westgate Center Drive, Hadley, MA 01035
http://northeast.fws.gov
Contact: Call field office

Region Six (CO, KS, MT, NE, ND, SD, UT, WY)

Division of Endangered Species
P.O. Box 25486, Denver Federal Center, Denver, CO 80225
http://mountain-prairie.fws.gov
Contact: Call field office

Region Seven (AK)

Division of Endangered Species
1011 E. Tudor Rd., Anchorage, AK 99503
http://alaska.fws.gov
Contact: Letter to Division of Endangered Species

APPENDIX H

GUIDANCE TO THE U.S. NUCLEAR REGULATORY COMMISSION STAFF ON THE RADIUM BENCHMARK DOSE APPROACH

H1.0 BACKGROUND

In 10 CFR 40.4, byproduct material is defined as the tailings or waste produced by the extraction or concentration of uranium or thorium from any ore processed primarily for its source material content, including discrete surface wastes resulting from uranium solution extraction processes. Uranium milling is defined as any activity resulting in byproduct material. Therefore, 10 CFR Part 40, Appendix A, applies to *in situ* leach, heap leach, and ion-exchange facilities that produce byproduct material, as well as to conventional uranium and thorium mills. This guidance only addresses uranium recovery facilities because there are no currently licensed or planned thorium mills.

The final rule, "Radiological Criteria for License Termination of Uranium Recovery Facilities," became effective on June 11, 1999, and added the following paragraph after the "radium in soil" criteria in Appendix A, Criterion 6(6):

> Byproduct material containing concentrations of radionuclides other than radium in soil, and surface activity on remaining structures, must not result in a total effective dose equivalent (TEDE) exceeding the dose from cleanup of radium contaminated soil to the above standard (benchmark dose), and must be at levels which are as low as is reasonably achievable. If more that one residual radionuclide is present in the same 100-square-meter area, the sum of the ratios for each radionuclide of concentration present to the concentration limit, will not exceed 1 (unity). A calculation of the peak potential annual TEDE within 1,000 years to the average member of the critical group that would result from applying the radium standard (not including radon) on the site, must be submitted for approval. The use of decommissioning plans with benchmark doses which exceed 100 mrem/yr, before application of as low as is reasonably achievable, requires the approval of the Commission after consideration of the recommendation of NRC staff. This requirement for dose criteria does not apply to sites that have decommissioning plans for soil and structures approved before June 11, 1999.

H2.0 RADIUM BENCHMARK DOSE APPROACH

The general requirements for a decommissioning plan, including verification of soil contamination cleanup, are addressed in Chapter 5.0 of this standard review plan. This appendix discusses the NRC staff evaluation of the radium benchmark dose approach, specifically dose modeling and its application to site cleanup activities that should be addressed in the decommissioning plan for those uranium recovery facilities licensed by the NRC and subject to the new requirements for cleanup of contaminated soil and buildings under 10 CFR Part 40, Appendix A, Criterion 6(6), as amended in 1999. The facilities that did not have an approved decommissioning plan at the time the rule became final are required to reduce residual radioactivity, that is, byproduct material, as defined by 10 CFR Part 40, to levels based on the potential dose, excluding radon, resulting from the application of the radium (Ra-226) standard at the site. This is referred to as the radium benchmark dose approach.

This guidance also applies to any revised decommissioning plan submitted for NRC review and approval, after the final rule was effective. However, if a subject licensee can demonstrate that no contaminated buildings will remain, and that soil thorium-230 (Th-230) does not exceed 5 pCi/g (above background) in the surface and 15 pCi/g in subsurface soil in any 100-square-meter area that meets the radium standard, and the natural uranium (U-nat, that is, U-238, U-234, and U-235) level is less than 5 pCi/g above background, radium benchmark dose modeling is not required. If future modeling with site-specific parameters for uranium recovery sites indicates that this is not a protective approach, the guidance will be revised. Therefore, it would be prudent for a uranium recovery licensee to consider the potential dose from any residual thorium and uranium.

The unity "rule" mentioned in the new paragraph of Criterion 6(6) applies to all licensed residual radionuclides. Therefore, if the ore (processed by the facility), tailings, or process fluid analyses indicate that elevated levels of Th-232 could exist in certain areas after cleanup for Ra-226, some verification samples in those areas should be analyzed for Th-232 or Ra-228. The thorium (Th-232) chain radionuclides (above local background levels) in milling waste would have soil cleanup criteria similar to the uranium chain radionuclides. The staff considers the EPA memorandum of February 12, 1998, (Directive No. 9200.4–25) concerning use of 40 CFR Part 192 soil criteria for Comprehensive Environmental Response, Compensation and Liability Act sites, an acceptable approach. This means that the Th-230 and Th-232 should be limited to the same concentration as their radium progeny with the 5 pCi/g (0.19 Bq/g) criterion applying to the sum of the radium constituents (Ra-226 plus Ra-228) as well as the sum of the thorium constituents (Th-230 plus Th-232) above background.

H2.1 Radium Benchmark Dose Modeling

H2.1.1 Areas of Review

The radium benchmark dose approach involves calculation of the peak potential dose for the site resulting from the 5 pCi/g [0.19 Bq/g] concentration of radium in the surface 15 cm [6 in.] of soil. The dose from the 15 pCi/g [0.56 Bq/g] subsurface radium would also be calculated for any area where the criterion is applied. The dose modeling review involves examining the computer code or other calculations employed for the dose estimates, the code or calculation input values and assumptions, and the modeling results (data presentation).

Evaluation of the radium benchmark dose modeling as proposed in the decommissioning plan, requires an understanding of the site conditions and site operations. The relevant site information presented in the plan or portions of previously submitted documents (e.g., environmental reports, license renewal applications, reclamation plan, and characterization survey report) should be reviewed.

H2.1.2 Review Procedures

The radium benchmark dose modeling review consists of ascertaining that an acceptable dose modeling computer code or other type of calculation has been used, that input parameter values appropriate (reasonable considering long-term conditions and representative of the application) for the site have been used in the modeling, that a realistic (overly conservative is not acceptable as it would result in higher allowable levels of uranium or thorium which would not be as low as is reasonably achievable) dose estimate is provided, and that the data presentation is clear and complete.

H2.1.3 Acceptance Criteria

The radium benchmark dose modeling results will be acceptable if the dose assessment (modeling) meets the following criteria:

(1)   Dose Modeling Codes and Calculations

The assumptions are considered reasonable for the site analysis, and the calculations employed are adequate. Reference to documentation concerning the code or calculations is provided [e.g., the RESRAD Handbook and Manual (Argonne National Laboratory, 1993a,b)].

The RESRAD code developed by the U.S. Department of Energy (Version 6.1, 2001) may be acceptable for dose calculations because, although the RESRAD ground-water calculations have limitations, this does not affect the uranium recovery sites that have deep aquifers (ground-water exposure pathway is insignificant). The DandD code developed for the NRC provides conservative default values, but does not, at this time, allow for modeling subsurface soil contamination and does not allow calculation of source removal due to soil erosion. The DandD code would not be adequate to model the dose from off-site contamination, but codes such as GENII are acceptable. See Appendix C of NUREG–1727 (NRC, 2000) for additional information.

If the code or calculations assumptions are not compatible with site conditions, adjustments have been made in the input to adequately reflect site conditions. For example, the RESRAD code assumes a circular contaminated zone. The shape factor (external gamma, code screen R017) is adjusted for an area that is not circular.

The code and/or calculation provides an estimated annual dose as total effective dose equivalent in mrem/yr. The DandD code provides the annual dose, but RESRAD calculates the highest instantaneous dose. However, RESRAD results are acceptable for long-lived radionuclides that do not move rapidly out of surface soils.

(2)   Input Parameter Values

The code/calculation input data are appropriate for the site and represent current or long-term conditions, whichever is more applicable to the time of maximum dose. When code default values are used, they are justified as appropriate (representative) for the site. Excessive conservatism (i.e., upper bound value) is not used, as this would result in a higher dose and thus higher levels of uranium and thorium could be allowed to remain on site.

Previously approved MILDOS code input parameter values may not be appropriate, because derived operational doses in the restricted area may be an order of magnitude higher than acceptable doses for areas to be released for unrestricted use.

Site-specific input values are demonstrated to be average values of an adequate sample size. Confidence limits are provided for important parameters so that the level of uncertainty can be estimated for that input value. Alteration of input values considers that some values are interrelated [see draft NUREG–1549, Appendix C (NRC, 1998)], and relevant parameters are modified accordingly. The preponderance of important parameter values are based on site measurements and not on conservative estimates. One or more models consider the annual average range of parameter values likely to occur within the next 200 years, for important parameters that can reasonably be estimated. Some other considerations for the input parameter values follow:

(a)   Scenarios for the Critical Group and Exposure Pathways

The scenario(s) chosen to model the potential dose to the average member of the critical group[1] from residual radionuclides at the site reflect reasonable probable future land use. The licensee has considered ranching, mining, home-based business, light industry, and residential farmer scenarios, and has justified the scenarios modeled.

On the basis of one or more of these projected (within 200 years is reasonably foreseeable) land uses to define the critical group(s), the licensee has determined and justified what exposure pathways are probable for potential exposure of the critical group to residual radionuclides at the site. Dairies are not likely to be established in the area of former uranium recovery facilities because the climate and soil restrict feed production. Even if some dairy cows were to graze in contaminated areas, the milk would probably be sent for processing (thus diluted), and not be consumed directly at the site. Therefore, milk consumption is not a likely ingestion exposure pathway. Also, a pond in the contaminated area providing a significant quantity of fish for the resident's diet is not likely, so the aquatic exposure pathway may not have to be modeled. However, the external gamma, plant ingestion, and inhalation pathways are likely to be important.

The radon pathway is excluded from the benchmark dose calculation as defined in Criterion 6(6) of Appendix A to 10 CFR Part 40. This also reflects the approach in the decommissioning rule (radiological criteria for license termination, 10 CFR Part 20, Subpart E).

---

[1]As defined in 10 CFR Part 20, "the group of individuals reasonably expected to receive the greatest exposure to residual radioactivity for any applicable set of circumstances."

(b)   Source Term

If the RESRAD code is used, the input includes lead-210 (Pb-210) at the same concentration input value as for Ra-226. The other radium progeny are automatically included in the code calculations. The chemical form of the contamination in the environment is considered in determining input values related to transport, or inhalation class (retention in the lung) for dose conversion factors.

(c)   Time Periods

The time periods for calculation of the dose from soil Ra-226 include the 1,000-year time frame. The calculated maximum annual dose and the year of occurrence are presented in the results.

(d)   Cover and Contaminated Zone

A cover depth of zero is used in the surface contamination model, and a depth of at least 15 cm [6 in.] is used for the subsurface model. The values for area and depth of contamination are derived from site characterization data. The erosion rate value for the contaminated zone is less than the RESRAD default value because in regions drier than normal, the erosion rate is less, as discussed in the RESRAD Data Collection Handbook (Argonne National Laboratory, 1993a), and the proposed value is justified. The soil properties are based on site data (sandy loam or sandy silty loam are typical for uranium recovery sites), and other input parameters are based on this demonstration of site soil type [see RESRAD handbook, pp., 23, 29, 77, and 105 (Argonne National Laboratory, 1993a)].

The evapotranspiration coefficient for the semi-arid uranium recovery sites is between 0.6 and 0.99. The precipitation value is based on annual values averaged over at least 20 years, obtained from the site or from a nearby meteorological station.

The irrigation rate value may be zero, or less than a code's default value, if supported by data on county or regional irrigation practices (e.g., zero is acceptable if irrigation water is obtained from a river not a well). The runoff coefficient value is based on the site's soil type, expected land use, and regional morphology.

(e)   Saturated Zone

The dry bulk density, porosity, "b" parameter, and hydraulic conductivity values are based on local soil properties. The hydraulic gradient for an unconfined aquifer is approximately the slope of the water table. For a confined aquifer, it represents the difference in potentiometric surfaces over a unit distance.

If the RESRAD code is used, the non-dispersion model parameter is chosen for areas greater than 1,000 square meters (code screen R014), and the well pump rate is based on irrigation, stock, or drinking water well pump rates in the area.

(f)   Uncontaminated and Unsaturated Strata

The thickness value represents the typical distance from the soil contamination to the saturated zone. Since the upper aquifer at uranium recovery sites is often of poor quality and quantity, the depth of the most shallow well used for irrigation or stock water in the region is chosen for the unsaturated zone thickness. A value of 18 m [60 ft] is typical for most sites {15 m [50 ft] for the Nebraska site}, but regional data are provided for justification. The density, porosity, and "b" parameter values are similar to those for the saturated zone, or any changes are justified.

(g)   Distribution Coefficients and Leach Rates

The distribution coefficient (Kd) is based on the physical and chemical characteristics of the soil at the site. The leach rate value of zero in the RESRAD code is acceptable as it allows calculation of the value. If a value greater than zero is given, the value is justified.

(h)   Inhalation

An average inhalation rate value of approximately 8,395 m^3/yr is used for the activity assumed for the rancher or farmer scenario based on a draft letter report (Sandia National Laboratories, 1998a). The mass loading for inhalation (air dust loading factor) value is justified based on the average level of airborne dust in the local region for similar activities as assumed in the model.

(i)   External Gamma

The shielding factor for gamma is in the range of 0.3 to 0.6 (70 to 40 percent shielding) (the DandD code screening default value is 0.55 and EPA has recommended 0.4). The factor is influenced by the type (foundation, materials) of structures likely to be built on the site and the gamma energy of the radionuclides under consideration.

The time fractions for indoor and outdoor occupancy are similar to default values in RESRAD and draft guidance developed for the decommissioning rule [NUREG/CR–5512, Volume 3 (NRC, 1999b)]. For example, the staff would consider fraction values approximating 0.7 indoors and 0.15 outdoors for a resident working at home, and 0.5 outdoors and 0.25 indoors for the farmer scenario (the remaining fraction allocated to time spent offsite).

The site-specific windspeed value is based on adequate site data. The average annual windspeed for the uranium recovery sites varies from 3.1 to

5.5 meters/sec [7 to 13 mph]. The maximum and annual average windspeed are also considered when evaluating proposed erosion rates.

(j) Ingestion

Average consumption values (g/yr) for the various types of foods are based on average values as discussed in NUREG/CR–5512, Volume 3 (NRC, 1999), or the Sandia Draft Letter Reports (Sandia National Laboratories,1998a,b), or are otherwise justified. Livestock ingestion parameters are default values, or are otherwise justified.

For sites with more than 100 acres of contamination, the fraction of diet from the contaminated area is assumed to be 0.25 for the farmer scenario (Sandia National Laboratories, 1998a), or is otherwise justified based on current or anticipated regional consumption practices for home-grown food. Because of the low level of precipitation in the areas in which uranium recovery facilities are located, extensive gardens or dense animal grazing is not likely, so the percentage of the diet obtained from contaminated areas would be lower than the code default value.

Note that the default plant mass loading factor in the DandD code can reasonably be reduced to 1 percent (Sandia National Laboratories, 1998c). The depth of roots is an important input parameter for uranium recovery licensees using the RESRAD code. The value is justified based on the type of crops likely to be grown on the site in the future. For vegetable gardens, a value of 0.3 is more appropriate than the RESRAD default value of 0.9 meters that is reasonable for alfalfa or for a similar deep-rooted plant.

(3) Presentation of Modeling Results

The radium benchmark dose modeling section of the decommissioning plan includes the code or calculation results as the maximum annual dose (total effective dose equivalent) in mrem/yr, the year that this dose would occur, and the major exposure pathways by percentage of total dose. The modeling section also includes discussion of the likelihood of the various land-use scenarios modeled (reflecting the probable critical groups), and provides the variations in dose (dose distribution) created by changing key parameter values to reflect the range of dose values that are likely to occur on the site in the future. The section also contains the results of a sensitivity analysis (RESRAD can provide a sensitivity analysis via the graphics function) to identify the important parameters for each scenario.

H2.1.4 Evaluation Findings

If the staff review, as described in this section, results in the acceptance of the radium benchmark dose modeling, the following conclusions may be presented in the technical evaluation report.

Appendix H

The staff has completed its review of the site benchmark dose modeling for the
_____ uranium mill facility. This review included an evaluation using the review
procedures and the acceptance criteria outlined in Section H2.1 of Appendix H of this standard
review plan.

The licensee has provided an acceptable radium benchmark dose model, and the staff
evaluation has determined that (1) the computer code or set of calculations used to model the
benchmark dose is appropriate for the site, (2) input parameter values used in each dose
assessment model are site-specific or reasonable estimates, and (3) the dose modeling results
include adequate estimates of dose uncertainty.

On the basis of the information presented in the application, and the detailed review conducted
of radium benchmark dose modeling for the _____ uranium mill facility, the staff
concludes that the information is acceptable and is in compliance with 10 CFR Part 40,
Appendix A, Criterion 6(6), which provides requirements for soil and structure cleanup.

H2.2 Implementation of the Benchmark Dose

H2.2.1 Areas of Review

The results of the radium benchmark dose calculations are used to establish a surface and
subsurface soil dose limit for residual radionuclides other than radium, as well as a limit for
surface activity on structures that will remain after decommissioning. The staff should review
the licensee's conversion of the benchmark dose limit to soil concentration (pCi/g) or surface
activity levels (dpm/100 cm^2) as a first step to determine cleanup levels. Alternatively, the
licensee can derive the estimated dose from the uranium or thorium contamination (as
discussed in Section 2.1.3) and compare this to the radium benchmark dose.

The reviewer should also evaluate the proposed cleanup guideline levels (derived concentration
limit) in relation to the "as low as is reasonably achievable" requirement and the unity rule.

H2.2.2 Review Procedures

The decommissioning plan section on cleanup criteria will be evaluated for appropriate
conversion of the radium standard benchmark dose to cleanup limits for soil uranium and
thorium and/or surface activity. The plan will also be examined to ensure reasonable
application of "as low as is reasonably achievable" to the cleanup guideline values and
application of the unity rule where appropriate.

H2.2.3 Acceptance Criteria

(1)   The soil concentration limit is derived from the site radium dose estimate. The modeling
       performed to estimate mrem/year per pCi/g of Th-230 and/or U-nat follows the criteria
       listed in Section 2.1.3. In addition, the U-nat source term input is represented as
       percent activity by 48.9 percent U-238, 48.9 percent U-234, and 2.2 percent U-235, or is
       based on analyses of the ore processed at the site. For a soil uranium criterion (derived

H–8

concentration limit), the chemical toxicity is considered in deriving a soil concentration limit if soluble forms of uranium are present.

(2)   Detailed justification for the inhalation pathway parameters is provided, such as the determination of the chemical form in the environment, to support the inhalation class.

(3)   The derived Th-230 soil limit will not cause any 100 square meter (m^2) area to exceed the Ra-226 limit at 1,000 years (i.e., current concentrations of Th-230 are less than 14 pCi/g surface and 43 pCi/g subsurface, if Ra-226 is at approximately background levels).

(4)   In conjunction with the activity limit, the "as low as is reasonably achievable" principle is considered in setting cleanup levels (derived concentration guideline levels). The as low as is reasonably achievable guidance in NUREG–1727, Appendix D (NRC, 2000) is considered. The proposed levels allow the licensee to demonstrate that the 10 CFR 40.42 (k) requirements (the premises are suitable for release, and reasonable effort has been made to eliminate residual radioactive contamination) can be met.

(5)   In recent practice at mill sites, the as low as is reasonably achievable principle is implemented by removing about 2 more inches [5 cm] of soil than is estimated to achieve the radium standard (reduce any possible excess or borderline contamination). At mills, it is generally cheaper to remove more soil than to do sampling and testing that may indicate failure and require additional soil removal with additional testing.

(6)   The unity rule is applied to the cleanup if more than one residual radionuclide is present in a soil verification grid (100 m^2). This means that the sum of the ratios for each radionuclide of the concentration present/concentration limit may not exceed 1 (i.e., unity).

(7)   The subsurface soil standard, if it is to be used, is applied to small areas of deep excavation where at least 15 cm [6 in.] of compacted clean fill is to be placed on the surface and where that depth of cover is expected to remain in place for the foreseeable future. The long-term cover depth used in the model is justified.

(8)   The surface activity limit for remaining structures is appropriately derived using an approved code or calculation. Because recent conservative dose modeling by NRC staff has indicated that more than 2,000 dpm/100 cm^2 alpha (U-nat or uranium chain radionuclides) in habitable buildings [2,000 hr/yr] could exceed an effective dose equivalent of 25 mrem/yr, the licensee proposes a total (fixed plus removable) average surface activity limit for such buildings that is lower than 2,000 dpm/100 cm^2, or a higher value is suitably justified.

(9)   If the DandD code is used, data are provided to support that 10 percent or less of the surface activity is removable; otherwise the resuspension factor is scaled to reflect the site-specific removable fraction. Note that this code assumes that the contamination is only on the floor, which can be overly conservative. If the RESRAD-Build code is used, the modeled distribution of contamination on walls and floor is justified.

Appendix H

H2.2.4 Evaluation Findings

If the staff review, as described in this section, results in the acceptance of the application of the radium benchmark dose modeling to the site cleanup criteria, the following conclusions may be presented in the technical evaluation report.

The staff has completed its review of the proposed implementation of the benchmark dose modeling results for the _____ uranium mill facility. This review included an evaluation using the review procedures and the acceptance criteria outlined in Section H2.2 of Appendix H of this standard review plan.

The licensee has provided an acceptable implementation plan of the benchmark dose modeling results to the proposed site cleanup activities, and the staff evaluation has determined that (1) the cleanup criteria will allow the licensee to meet 10 CFR Part 40.42(k) and 10 CFR Part 40, Appendix A, Criterion 6(6) requirements; (2) the soil and structures of the decommissioned site will permit termination of the license because public health and the environment will not be adversely affected by any residual radionuclides.

H3.0 REFERENCES

Argonne National Laboratory. "Data Collection Handbook to Support Modeling the Impacts of Radioactive Material in Soil." ANL/EAIS–8. Washington, DC: U.S. Department of Energy. April 1993a.

———. "Manual for Implementing Residual Radioactive Material Guidelines Using RESRAD, Version 5.0." ANL/EAD/LD–2. Washington, DC: U.S. Department of Energy. September 1993b.

NRC. NUREG/CR–5512, "Residual Radioactive Contamination from Decommissioning—User's Manual." Vol. 2. Washington, DC: NRC. April 2001.

———. NUREG–1727, "NMSS Decommissioning Standard Review Plan." Washington, DC: NRC. September 2000.

———. NUREG/CR–5512, "Residual Radioactive Contamination from Decommissioning—Parameter Analysis." Draft. Vol. 3. Washington, DC: NRC. October 1999.

———. Draft NUREG–1549, "Decision Methods for Dose Assessment to Comply With Radiological Criteria for License Termination." Washington, DC: NRC. July 1998.

Sandia National Laboratories. "Review of Parameter Data for the NUREG/CR–5512 Residential Farmer Scenario and Probability Distributions for the DandD Parameter Analysis." Draft Letter Report. January 30, 1998a.

———. "Review of Parameter Data for the NUREG/CR-5512 Building Occupancy Scenario and Probability Distributions for the DandD Parameter Analysis." Draft Letter Report. January 30, 1998b.

———. "Comparison of the Models and Assumptions Used in the DandD 1.0, RESRAD 5.61, and RESRAD-Build Computer Codes with Respect to the Residential Farmer and Industrial Occupant Scenarios Provided in NUREG/CR–5512." Draft Report. October 15, 1998c.

APPENDIX I

REGULATORY ISSUE SUMMARY 2000-23

In Regulatory Issue Summary 2000–23 (NRC,2000), the Commission promulgated staff requirements to revise previously published guidance on matters related to regulation of uranium recovery facilities. This appendix presents that revised guidance as it relates to uranium mills and is included to assist the staff and operators when considering related actions. The issues and policies addressed are:

(1)   Part 41 Rulemaking (SECY–99–011)

(2)   Disposal of Non-11e.(2) Byproduct Material in Tailings Impoundments (SECY–99–012)

(3)   Processing of Material Other Than Natural Uranium Ores (SECY–99–012)

(4)   Concurrent Jurisdiction of Non-Radiological Hazards of Uranium Mill Tailings (SECY–99–277)

Reference

NRC. Regulatory Issue Summary 2000-23. "Recent Changes to Uranium Recovery Policy." Washington, DC: NRC. November 2000.

Appendix I

Recent Changes to Uranium Recovery Policy

November 30, 2000

- ADDRESSEES
- INTENT
- BACKGROUND
- PART 41 RULEMAKING (SECY-99-011)
- DISPOSAL OF NON-11e.(2) BYPRODUCT MATERIAL IN TAILINGS IMPOUNDMENTS (SECY-99-012)
- PROCESSING OF MATERIAL OTHER THAN NATURAL URANIUM ORES (SECY-99-012)
- CLASSIFICATION OF LIQUID WASTES AT ISL FACILITIES (SECY-99-013)
- GROUND-WATER ISSUES AT ISL FACILITIES (SECY-99-013)
- CONCURRENT JURISDICTION OF NON-RADIOLOGICAL HAZARDS OF URANIUM MILL TAILINGS (SECY-99-277)
- SUMMARY OF ISSUES

ADDRESSEES

All holders of materials licenses for uranium and thorium recovery facilities.

INTENT

The U.S. Nuclear Regulatory Commission (NRC) is issuing this regulatory issue summary (RIS) to inform materials licensees of the Commission's decisions on four Commission Papers prepared by the Uranium Recovery staff and the Office of the General Counsel (OGC). All the policy decisions will be codified in the 10 CFR Part 41 rulemaking that has been initiated. No specific action nor written response is required.

BACKGROUND

NRC staff prepared four Commission Papers in 1999 to address various uranium recovery issues. One Commission Paper (SECY-99-011, "Draft Rulemaking Plan; Domestic Licensing of Uranium and Thorium Recovery facilities—Proposed New 10 CFR Part 41") addressed the need to revise and update uranium recovery regulations, particularly with respect to in situ leach (ISL) facilities and recommended the initiation of rulemaking to create a new Part 41 specific to uranium recovery. The other three Commission Papers addressed issues raised by the National Mining Association (NMA) in its April 1998 paper, "Recommendations for a Coordinated Approach to Regulating the Uranium Recovery Industry." The first of those papers (SECY-99-012, "Use of Uranium Mill Tailings Impoundments for the Disposal of Other Than 11e(2) Byproduct Materials, and Reviews of Applications to Process Material Other Than Natural Ore") discussed the disposal of radioactive waste, other than byproduct material, defined in section 11e.(2) of the Atomic Energy Act (AEA) of 1954, as amended, in mill tailings impoundments, and the processing of material, other than natural ore, for source material at licensed uranium mills. The second of those papers (SECY-99-013, "Recommendations on ways to Improve the Efficiency of NRC Regulation at *In Situ* Leach Uranium Recovery Facilities") discussed the regulation of ground water at ISL sites and the issue of which waste streams at ISL facilities come under NRC regulatory jurisdiction as 11e.(2) byproduct material. The last paper (SECY–99–277, "Concurrent Jurisdiction of Non-Radiological Hazards of Uranium Mill Tailings) addressed the issue of concurrent jurisdiction (with States that do not have Agreement State regulatory authority for 11e.(2) material under section 274 of the AEA) over the non-radiological hazards of uranium mill tailings.

On July 13, 2000, the Commission issued a Staff Requirements Memorandum (SRM) on SECY–99–011. On July 26, 2000, the Commission issued SRMs on SECY–99–012 and SECY–99–013, and on August 11, 2000, the SRM on SECY–99–277 was issued.

The decisions and directions in these SRMs and the staff actions in response are discussed in sections that follow.

PART 41 RULEMAKING (SECY–99–011)

SECY–99–011 approved the staff's recommendation to provide a draft Rulemaking Plan (RP) for comment to the Agreement States, with the preferred option being the creation of a new Part 41 dedicated to uranium recovery regulation. The Commission directed the staff to revise the draft RP to reflect the Commission's guidance in the other uranium recovery SRMs.

On September 11, 2000, the staff transmitted the draft RP to all States for comment. The staff sent the draft RP to all States rather than just Agreement States because the issue of concurrent jurisdiction regarding non-radiological hazards primarily affects non-Agreement States, and the staff wanted to give those States an opportunity to comment on the draft RP. Comments have been received from several States. In addition, the NMA and two licensees provided comments on the draft RP. The staff will consider all the comments received in preparing its final RP, which it expects to issue in early 2001.

DISPOSAL OF NON-11e.(2) BYPRODUCT MATERIAL IN TAILINGS IMPOUNDMENTS (SECY–99–012)

In 1995, the staff published guidance, in the Federal Register [exit icon] (60 FR 49296), for the disposal, in uranium mill tailings impoundments, of radioactive material that is not byproduct material, as defined in section 11e.(2) of the AEA. The guidance consisted of 10 criteria to determine whether to approve a proposed disposal of non-11e.(2) byproduct material in a uranium mill tailings impoundment. In its 1998 white paper, the NMA emphasized that the criteria were too restrictive, pointing out that no requests for such disposals have been made since the guidance was issued. The Commission, in the SRM for SECY–99–012, approved an option that would allow more flexibility in permitting non-11e.(2) material to be disposed of in tailings impoundments. The NRC intends to incorporate the criteria into the new Part 41. In the interim, the Commission directed the staff to implement the SRM.

To comply with the direction in the SRM, the staff is revising the 1995 guidance in the following manner:

- The staff will remove the prohibitions, found in items 2, 4, and 5, regarding non-AEA radioactive material and material subject to regulation under other legislative authorities, such as the Toxic Substance Control Act (TSCA) or the Resource Conservation and Recovery Act (RCRA).

- The staff will add a criterion regarding approval from the appropriate regulators of TSCA, RCRA, and non-AEA radioactive material for disposal of such material in the tailings impoundment.

- The staff will revise the criterion, in item 8, regarding approval by Low-Level Waste Compacts, to allow for the situation in which material proposed for disposal does not fall under the jurisdiction of Low-Level Waste Compacts (e.g., radioactive material not regulated under the AEA).

- The Commission directed the staff to pursue a generic exemption to NRC's disposal requirements for low-level radioactive waste in 10 CFR Part 61, rather than having to grant an exemption, under 10 CFR 61.6, as identified in item 10. A generic exemption to regulations must be issued through a rulemaking process. Therefore, the staff will pursue incorporating the generic exemption in the new Part 41. In the interim, the requirement for a specific exemption will remain in the guidance, with addition of a caveat for material not regulated under Part 61.

The staff therefore is revising its 1995 guidance. The complete revised guidance, is in Attachment 1.

PROCESSING OF MATERIAL OTHER THAN NATURAL URANIUM ORES (SECY-99-012)

In 1995, the staff published its position and guidance, in the Federal Register (60 FR 49296), on the use of uranium feed material other than natural ores (alternate feed material), in uranium mills. The guidance identified three determinations that the staff had to make in order to approve an alternate feed request. The third determination—whether the ore is being processed primarily for its source material content—generated considerable controversy. This determination was required to address the concern that wastes that would otherwise have to be disposed of as radioactive or mixed waste would be proposed for processing at a uranium mill primarily to be able to dispose of them in the tailings pile as 11e.(2) byproduct material. This determination was essentially a determination of the motives of the mill operator in requesting approval of a specific stream of alternate feed material. In many cases it involved questioning the financial aspects of acquiring and processing the alternate feed material, and selling the resultant uranium product.

In its 1998 white paper, the NMA emphasized that NRC should not be looking to a licensee's motives in processing alternate feed material. After careful consideration of stakeholder comments and the staff's analysis, the Commission, in the SRM for SECY–99–012, directed the staff to allow processing of alternate feed material without inquiry into a licencee's economic motives, and referred to a Commission decision (CLI–00–01 51 NRC 9) on a specific instance of proposed processing of alternate feed, that was brought before the Atomic Safety Licensing Board and then appealed to the Commission. The Commission also addressed the second determination in the 1995 guidance (i.e., whether the feed material contains hazardous waste). It directed the staff to allow more flexibility with regard to this issue consistent with its direction to the staff on the disposal of non-11e.(2) byproduct material in tailings piles.

The Commission directed the staff to revise, issue, and implement final guidance on the processing of alternate feed as soon as possible and to codify the guidance in the new Part 41.

To comply with the SRM, the staff is revising the 1995 position and guidance in the following manner:

The staff will modify the prohibition in item 2 on feed material containing hazardous waste, to allow such feed material provided that the licensee obtains approval of the U.S. Environmental Protection Agency (EPA) or the State, and a commitment from the long-term custodian to accept the tailings after site closure.

The staff will revise the manner in which it determines whether the ore is being processed primarily for its source material content, to focus on the product of the processing, and eliminate any inquiry into the licensee's economic motives for the processing.

The staff therefore is revising its 1995 guidance. The complete revised guidance, is in Attachment 2.

CLASSIFICATION OF LIQUID WASTES AT ISL FACILITIES (SECY–99–013)

Before 1995, the staff practice for addressing the disposal of evaporation pond sludges at ISL facilities relied on a broad reading of the definition of 11e.(2) byproduct material. This broad reading only addressed discrete surface wastes capable of controlled disposal and did not distinguish between wastes generated at various phases of an ISL operation. All waste materials generated during ISL operations and ground-water restoration activities were designated 11e.(2) byproduct material and disposed of at licensed uranium mill tailings impoundments, in accordance with 10 CFR Part 40, Appendix A, Criterion 2.

The staff issued two guidance documents in 1995 to address issues raised by the industry in the uranium recovery program. The first, "Staff Technical Position on Effluent Disposal at Licensed Uranium Recovery Facilities" (hereinafter, the effluent guidance), was intended to ensure protection of the environment and public, while providing uranium recovery licensees with flexibility regarding the disposal of various types of liquid effluents generated during the operation of their facilities. In issuing this guidance, the staff took a more narrow view of the definition of 11e.(2) byproduct material. It differentiated between the various waste waters generated during ISL operations on the basis of their origin and whether uranium was extracted for its source material content during that phase of the operation. Waste waters and the associated solids produced during the uranium extraction phase of site operations, called "production bleed," were classified as AEA Section 11e.(2) byproduct material and therefore subject to regulation by NRC. Conversely, waste waters and the resulting solids produced after uranium extraction (i.e., during ground-water restoration activities) were classified as "mine waste waters," and therefore were subject to regulation by individual States under their applicable mining programs. These wastes were considered naturally occurring radioactive material (NORM). However, because licensees often dispose of waste waters from uranium extraction and post-extraction activities in the same evaporation ponds, the resulting solids are a commingled waste consisting of 11e.(2) byproduct material and sludges derived from mine waste water.

In the second guidance document, "Final Revised Guidance on Disposal of Non-Atomic Energy Act of 1954, Section 11e.(2) Byproduct Material in Tailings Impoundments" (hereinafter, the disposal guidance), the staff identified 10 criteria that licensees should meet before NRC could authorize the disposal of AEA material other than 11e.(2) byproduct material in tailings impoundments. One of these criteria prohibited the disposal of radioactive material not covered by the AEA, including NORM (see earlier discussion for policy revisions). This criterion was intended to avoid the possibility of dual regulation of the radioactive constituents in the impoundments, since individual States are responsible for radioactive materials not covered by the AEA.

The industry expressed concerns, in NMA's white paper, that, taken together, these two guidance documents leave no option for the disposal of radioactively contaminated sludges from ISL evaporation ponds. The reason for this concern is that the 11e.(2) byproduct material was commingled with a NORM waste, which the disposal guidance prohibits from disposal in a tailings impoundment. The industry emphasized that the staff's waste classification, based on the origin of the waste water (i.e., from the extraction or restoration phase) at an ISL facility, makes the disposal of such sludges in a mill tailings impoundment, as required under Criterion 2 of 10 CFR Part 40, Appendix A, impossible—even though the sludges derived from waste waters produced throughout a facility's life cycle are physically, chemically, and radiologically identical.

The staff analyzed several options in SECY-99-013 for addressing the industry's concerns. In the SRM for SECY-99-013, the Commission determined that all liquid effluents at ISL uranium recovery facilities are 11e.(2) byproduct material. NRC takes the position that any waste water generated during or after the uranium extraction phase of site operations, and all evaporation pond sludges derived from such waste waters, are classified as 11e.(2) byproduct material. The staff will make no legal distinction among the waste waters produced at different stages in a facility's life cycle.

This revised policy is effective immediately. The staff intends to codify this policy in the new rulemaking for Part 41 and associated regulatory guidance.

GROUND-WATER ISSUES AT ISL FACILITIES (SECY-99-013)

Over the past several years, the industry has expressed concern that NRC's regulation of ground water at ISLs is duplicative of the ground-water protection programs required by the Safe Drinking Water Act (SDWA), as administered by EPA or EPA-authorized States. EPA and the States protect ground-water quality through the Underground Injection Control (UIC) program, under the SDWA. The States often require additional measures in the UIC program that are more stringent than the Federal program. As presented in NMA's white paper, the industry contended that NRC's review and licensing activities are a duplicative form of regulation covering the same issues. Additionally, NMA also expressed the view that NRC did not have authority to regulate ground water at ISLs.

Historically, NRC has imposed conditions on ISL operations to ensure that ground-water quality is maintained during licensed activities and that actions are taken to ensure the restoration of ground-water quality before the license is terminated. The specific conditions imposed in an ISL license have typically been the result of NRC's independent review, as documented in safety evaluation reports and appropriate environmental evaluations.

In addition to NRC's review, licensees must also obtain a UIC permit from EPA or the EPA-authorized State before uranium recovery operations can begin. EPA or the authorized State conducts many of the same types of reviews as NRC. This is evidenced by NRC incorporating ground-water protection limits from a State's permitting program into specific license requirements, after conducting its own review of the licensee's groundwater protection program, including the use of State-imposed standards—and staff routinely accepting specific methodologies and guidance developed by EPA or States for ground-water monitoring programs and well construction.

In the SRM for SECY–99–013, the Commission approved the staff continuing discussions with EPA and appropriate States to determine the extent to which NRC can rely on the EPA UIC program for ground-water protection issues, thereby potentially minimizing duplicative review of ground-water protection at ISL facilities. Part of the discussions with EPA and appropriate States should include appropriate methods to implement any agreements, including Memoranda of Understanding (if necessary) and potential requirements that could be incorporated in the new Part 41. In the interim, it is recognized that some NRC/EPA dual regulation of the ground-water at ISL facilities will continue until such time that NRC can defer to EPA's UIC program.

NRC has initiated a new round of discussions with the EPA since the Commission decision in July 2000, and discussions with the appropriate States should begin in early to mid 2001.

In February 1998, staff documented its review process for ISLs, including a detailed evaluation of ground-water activities, in a draft Standard Review Plan (draft SRP) for ISL facility license applications (NUREG–1569), that was published for public comment. Following the comment period, staff held a public workshop on the SRP to discuss the issues raised. The staff intends to use the draft SRP in licensing reviews until the rulemaking for new Part 41 (SECY–99–011) has been completed and NUREG–1569 is finalized.

CONCURRENT JURISDICTION OF NON-RADIOLOGICAL HAZARDS OF URANIUM MILL TAILINGS (SECY–99–277)

In 1980, the staff considered the issue of whether the Uranium Mill Tailings Radiation Control Act (UMTRCA) preempts a non-Agreement State's authority to regulate the non-radiological hazards associated with 11e.(2) byproduct material and concluded that it did not. The NRC concluded that NRC and the State both exercised this authority. As a result, the staff has

Appendix I

followed the practice of sharing jurisdiction of the non-radiological hazards with States. In its 1998 white paper, the NMA questioned the 1980 staff interpretation of UMTRCA. The Commission, in the SRM for SECY–99–0277 determined that NRC has exclusive jurisdiction over both the radiological and non-radiological hazards of 11e.(2) byproduct material.

As a result of this decision, the staff will implement its exclusive authority over the non-radiological hazards of 11e.(2) byproduct material and not recognize State authority in this area.

SUMMARY OF ISSUES

The Commission has evaluated a range of uranium recovery issues and the staff evaluation and has directed, through SRMs, the staff to take various actions that will ultimately be incorporated into the new Part 41 rulemaking and existing uranium recovery SRPs.

In the interim, this RIS informs the licensees of the Commission's decisions. These are: (1) to allow more flexibility in the disposal of non-11e.(2) material in tailings impoundments, subject to certain considerations; (2) to allow alternate feed material to be processed for uranium (or thorium) without any inquiry into a licensee's economic motives; (3) to classify all waste water and sludges generated during or after the uranium (or thorium) extraction phase of *in situ* leach operations as 11e.(2) byproduct material; (4) to continue discussions with EPA and appropriate States to determine the extent that NRC can rely on the EPA UIC program for ground-water protection at ISL facilities; and (5) to note that NRC has exclusive jurisdiction over both the radiological and non-radiological hazards of 11e.(2) byproduct material.

This regulatory issue summary requires no specific action nor written response. If you have any questions about this summary, please contact the technical contact listed below.

Michael F. Weber, Director
Division of Fuel Cycle Safety & Safeguards
Office of Nuclear Material Safety and Safeguards

Technical Contact:

Kenneth R. Hooks, NMSS
301-415-7777
E-mail: krh1@nrc.gov

Attachments:

1. Interim Guidance Non-11e.(2)
2. Interim Position Alternate Feed
3. List of Recently Issued NRC Regulatory Issue Summaries

(ADAMS Accession Number ML003773008)

ATTACHMENT 1

Interim Guidance on Disposal of Non-Atomic Energy Act of 1954, Section 11e.(2) Byproduct Material in Tailings Impoundments

1.   In reviewing licensee requests for the disposal of wastes that have radiological characteristics comparable to those of Atomic Energy Act of 1954, Section 11e.(2) byproduct material [hereafter designated as "11e.(2) byproduct material"] in tailings impoundments, the Nuclear Regulatory Commission staff will follow the guidance set forth below. Since mill tailings impoundments are already regulated under 10 CFR Part 40, licensing of the receipt and disposal of such material [hereafter designated as "non-11e.(2) byproduct material"] should also be done under 10 CFR Part 40.

2.   Special nuclear material and Section 11e.(1) byproduct material waste should not be considered as candidates for disposal in a tailings impoundment, without compelling reasons to the contrary. If staff believes that such material should be disposed of in a tailings impoundment in a specific instance, a request for Commission approval should be prepared.

3.   The 11e.(2) licensee must provide documentation showing necessary approvals of other affected regulators (e.g., the U.S. Environmental Protection Agency or State) for material containing listed hazardous wastes or any other material regulated by another Federal agency or State because of environmental or safety considerations.

4.   The 11e.(2) licensee must demonstrate that there will be no significant environmental impact from disposing of this material.

5.   The 11e.(2) licensee must demonstrate that the proposed disposal will not compromise the reclamation of the tailings impoundment by demonstrating compliance with the reclamation and closure criteria of Appendix A of 10 CFR Part 40.

6.   The 11e.(2) licensee must provide documentation showing approval by the Regional Low-Level Waste Compact in whose jurisdiction the waste originates as well as approval by the Compact in whose jurisdiction the disposal site is located, for material which otherwise would fall under Compact jurisdiction.

7.   The U.S. Department of Energy (DOE) and the State in which the tailings impoundment is located, should be informed of the U.S. Nuclear Regulatory Commission findings and proposed action, with a request to concur within 120 days. A concurrence and commitment from either DOE or the State to take title to the tailings impoundment after closure must be received before granting the license amendment to the 11e.(2) licensee.

8.   The mechanism to authorize the disposal of non-11e.(2) byproduct material in a tailings impoundment is an amendment to the mill license under 10 CFR Part 40, authorizing the receipt of the material and its disposal. Additionally, an exemption to the

Appendix I

requirements of 10 CFR Part 61, under the authority of 10 CFR 61.6, must be granted, if the material would otherwise be regulated under Part 61. (If the tailings impoundment is located in an Agreement State with low-level waste licensing authority, the State must take appropriate action to exempt the non-11e.(2) byproduct material from regulation as low-level waste.) The license amendment and the 10 CFR 61.6 exemption should be supported with a staff analysis addressing the issues discussed in this guidance.

ATTACHMENT 2

Interim Position and Guidance on the Use of Uranium Mill Feed Material Other Than Natural Ores

In reviewing licensee requests to process alternate feed material (material other than natural ore) in uranium mills, the Nuclear Regulatory Commission staff will follow the guidance presented below. Besides reviewing to determine compliance with appropriate aspects of Appendix A of 10 CFR Part 40, the staff should also address the following issues:

1. Determination of whether the feed material is ore.

For the tailings and wastes from the proposed processing to qualify as 11e.(2) byproduct material, the feed material must qualify as "ore." In determining whether the feed material is ore, the following definition of ore will be used:

Ore is a natural or native matter that may be mined and treated for the extraction of any of its constituents or any other matter from which source material is extracted in a licensed uranium or thorium mill.

2. Determination of whether the feed material contains hazardous waste.

If the proposed feed material contains hazardous waste, listed under subpart D Sections 261.30-33 of 40 CFR (or comparable Resource Conservation and Recovery Act (RCRA) authorized State regulations), it would be subject to the U.S. Environmental Protection Agency (EPA) or State regulation under RCRA. If the licensee can show that the proposed feed material does not contain a listed hazardous waste, this issue is resolved.

Feed material exhibiting only a characteristic of hazardous waste (ignitable, corrosive, reactive, toxic) would not be regulated as hazardous waste and could therefore be approved for recycling and extraction of source material. However, this does not apply to residues from water treatment, so determination that such residues are not subject to regulation under RCRA will depend on their not containing any characteristic hazardous waste. Staff may consult with EPA (or the State) before making a determination of whether the feed material contains hazardous waste.

If the feed material contains hazardous waste, the licensee can process it only if it obtains EPA (or State) approval and provides the necessary documentation to that effect. Additionally, for feed material containing hazardous waste, the staff will review documentation from the licensee that provides a commitment from the U.S. Department of Energy or the State to take title to the tailings impoundment after closure.

3. Determination of whether the ore is being processed primarily for its source-material content.

For the tailings and waste from the proposed processing to qualify as 11e.(2) byproduct material, the ore must be processed primarily for its source-material content. If the only product produced in the processing of the alternate feed is uranium product, this determination is satisfied. If, in addition to uranium product, another material is also produced in the processing of the ore, the licensee must provide documentation showing that the uranium product is the primary product produced.

If it can be determined, using the aforementioned guidance, that the proposed feed material meets the definition of ore, that it will not introduce a hazardous waste not otherwise exempted, or if it has been approved by the EPA (or State) and the long-term custodian, and that the primary purpose of its processing is for its source-material content, the request can be approved.

APPENDIX J

TECHNICAL EVALUATION OF APPENDIX A CRITERIA

During the review process, U.S. Nuclear Regulatory Commission (NRC) staff will verify that specific criteria of 10 CFR Part 40, Appendix A have been met. It is suggested that the technical reviewer prepare a list of the specific technical criteria and the method or design used to meet these criteria to be included in the technical evaluation report. The example offered shows one method of documentation.

J1.0 EXAMPLE OF TECHNICAL EVALUATION OF APPENDIX A CRITERIA

The following text is from an NRC technical evaluation report for a uranium mill facility, and represents the type of conclusions related to meeting specific technical criteria in 10 CFR Part 40, Appendix A.

CONCLUSIONS RELATED TO MEETING APPENDIX A CRITERIA

The staff further concludes that the specific criteria of 10 CFR Part 40 Appendix A are met as follows.

J1.1 Criterion 1

Demonstrate that erosion, disturbance, and dispersion by natural forces over the long term are minimized.

The contaminated tailings will be protected from flooding and erosion by an engineered rock riprap layer. The riprap has been designed in accordance with the guidance suggested by the NRC staff (NRC, 1990). The staff considers that erosion protection that meets that guidance will provide adequate protection against erosion and dispersion by natural forces over the long term. As discussed in technical evaluation report Sections 4.3 and 4.5, adequate protection is provided by (1) selection of proper rainfall and flooding events, (2) selection of appropriate parameters for determining flood discharges, (3) computation of flood discharges using appropriate and/or conservative methods, (4) computation of appropriate flood levels and flood forces associated with the design discharge, (5) use of appropriate methods for determining erosion protection needed to resist the forces produced by the design discharge, (6) selection of a rock type for the riprap layer that will be durable and capable of providing the necessary erosion protection for a long period of time, and (7) placement of a riprap layer in accordance with accepted engineering practice and in accordance with appropriate testing and quality assurance controls.

Demonstrate that the tailings are disposed of in a manner that does not require active maintenance to preserve conditions at the site.

As discussed in technical evaluation report Sections 4.3 and 4.5, the staff considers that the riprap layers proposed will not require active maintenance over the 1,000-year design life, for the following reasons: (1) the riprap has been designed to protect the tailings from rainfall and flooding events which have very low probabilities of occurrence over a 1,000-year period, resulting in no damage to the layers from those rare events; (2) the rock proposed for the riprap layers is designed to be durable and is not expected to deteriorate significantly over the with 1,000-year design life; and (3) during construction, the rock layers will be placed in accordance with appropriate engineering and testing practices, minimizing the potential for damage, dispersion, and segregation of the rock.

J1.2   Criterion 4

Demonstrate that the upstream rainfall catchment areas are minimized to decrease erosion potential and the size of the floods that could erode or wash out sections of the tailings disposal area.

The site is located in an area that is flooded by off-site floods from Moab Wash and the Colorado River. However, as discussed in the technical evaluation report, the site is protected from direct on-site precipitation and flooding by engineered riprap layers for the top and side slopes; the tailings disposal cell will need this protection regardless of where it is located. The riprap for the side slopes and drainage ditches is large enough to resist flooding from the minimal flow velocities of floods occurring from a probable maximum flood on the Colorado River. A large rock apron has been provided to provide protection against the potential migration of Moab Wash and the Colorado River. The staff therefore concludes that the erosion potential at the site has been acceptably minimized, since any flooding at the site is mitigated by the erosion protection, and the forces associated with off-site floods are minimal.

Demonstrate that topographic features provide good wind protection.

The staff considers that the site is adequately protected from wind erosion by the placement of an engineered riprap layer that protects the tailings from surface water erosion. Studies performed for the NRC staff have shown that an engineered riprap layer designed to protect against water erosion will be capable of providing adequate protection against wind erosion.

Demonstrate that embankments and cover slopes are relatively flat after stabilization to minimize erosion potential and to provide conservative factors of safety assuring long-term stability.

The relatively flat top and side slopes of the covers will be protected from erosion by an engineered riprap layer which is designed to provide long-term stability (technical evaluation report Section 4.3). The erosion potential of the covers is minimized by the designing the rock to be sufficiently large to resist flooding and erosion, based on the slope selected. Thus, the staff concludes that the slopes, with their corresponding rock designs, are sufficiently flat to meet this criterion.

Demonstrate that the rock cover reduces wind and water erosion to negligible levels, including consideration of such factors as the shape, size, composition, and gradation of the rock particles; rock cover thickness and zoning of particle size; and steepness of underlying slopes. Demonstrate that rock fragments are dense, sound, and resistant to abrasion, and free from cracks, seams, and other defects.

The contaminated tailings will be protected from flooding and erosion by an engineered rock riprap layer. The riprap has been designed in accordance with the guidance suggested by the NRC staff (NRC, 1990). As discussed in Sections 4.3 and 4.5 of the technical evaluation report, the staff considers that erosion protection which meets that guidance will provide adequate protection against erosion and dispersion by natural forces over the long term. Adequate protection is provided by (1) selection of proper rainfall and flooding events, (2) selection of appropriate parameters for determining flood discharges, (3) computation of flood discharges using appropriate and/or conservative methods, (4) computation of appropriate flood levels and flood forces associated with the design discharge, (5) use of appropriate methods for determining erosion protection needed to resist the forces produced by the design discharge, (6) selection of a rock type for the riprap layer that will be durable and capable of providing the necessary erosion protection for a long period of time, and (7) placement of a riprap layer in accordance with accepted engineering practice and in accordance with appropriate testing and quality assurance controls.

J1.3   Criterion 12

Demonstrate that active on-going maintenance is not necessary to preserve isolation.

As discussed in Sections 4.3 and 4.5 of the technical evaluation report, the staff considers that the erosion protection will not require active maintenance over the 1,000-year design life, for the following reasons: (1) the riprap has been designed to protect the tailings from rainfall and flooding events which have low probabilities of occurrence over a 1,000-year period, resulting in no damage to the layers from those rare events; (2) the rock proposed for the riprap layers is designed to be durable and is not expected to deteriorate significantly over the 1,000-year design life; and (3) during construction, the rock layers will be placed in accordance with appropriate engineering and testing practices, minimizing the potential for damage, dispersion, and segregation of the rock.

APPENDIX K

CONTENT AND FORMAT FOR ALTERNATE CONCENTRATION LIMIT APPLICATIONS

Application Content

The application should contain sufficient information to show that a hazardous constituent will not pose a substantial present or potential harm to human health or the environment, as long as the proposed Alternate Concentration Limit is not exceeded; and the proposed Alternate Concentration Limit is as low as reasonably achievable, considering practicable corrective actions. This demonstration should assess the hazards of the constituent in question and evaluate the consequences presented by potential exposures to the constituent. The application must consider the 19 factors listed in 10 CFR Part 40, Appendix A, Criterion 5B(6).

For ease of review, the application should address these factors through the following assessments. The Hazard Assessment should evaluate the radiological dose and toxicity of the constituents in question and the risk to human health and the environment posed by the constituents. The Exposure Assessment should examine the existing distribution of hazardous constituents, as well as the potential source(s) for future constituent releases. This should include the fate and transport of the hazardous constituents in ground water and hydraulically-connected surface water; and the potential consequences associated with human and environmental exposure to the hazardous constituents. The Corrective Action Assessment should (1) identify all realistic corrective action scenarios available; (2) assess their technical feasibility; (3) determine the costs and benefits associated with each scenario; and (4) select a practicable corrective action to achieve the hazardous constituent concentration that is protective of human health and the environment. The outcome of this assessment is a determination that the selected corrective action is as low as reasonably achievable.

There should be enough detailed information in the application to allow the NRC reviewer to independently verify that the proposed Alternate Concentration Limit will not pose a significant present or future hazard to human health or the environment, and that the limit is as low as reasonably achievable, considering practicable corrective actions. Site characteristics, milling processes, disposal operations, and ore composition should be discussed in the application. Information related to each of the 19 factors listed in 10 CFR Part 40, Appendix A, Criterion 5B(6) should be addressed; however, all of these factors may not be applicable to every site. If this is the case, the applicant must explain why a particular factor is not appropriate. For example, ground-water discharging into surface waters may not occur near a mill tailings site. Therefore, stream flow characteristics and transport assessments within the surface water may not be necessary. In any regard, the burden of proof resides with the applicant to demonstrate that selected factors do not need to be considered.

Much of this detailed information may already be available in existing licensing documents, such as environmental reports, license applications, or annual compliance monitoring reports. This information can be readily incorporated into the Alternate Concentration Limit application to produce a stand-alone document. The applicant may simply reference this existing information; however, additional time and NRC resources will be needed to collect the information from the licensee's docket file in order to proceed with the detailed review.

Appendix K

Application Format

A standard application format helps to assure the application contains information that addresses all applicable regulatory requirements, and helps to guide both the reviewer and interested stakeholders to pertinent and crucial information. A standard format also greatly contributes to the time efficiency and effectiveness of the review process. The applicant is not required to follow this standard format. Any application, regardless of format, that adequately addresses the suitability of a proposed Alternate Concentration Limit is acceptable for NRC review; however, reviewing an application with a significantly different format will likely require considerably more NRC staff time and resources to conduct the review. An applicant is strongly encouraged to provide a cross-reference table comparing the standard format to the format used in the application, if that format significantly differs from the standard format.

The applicant should present the technical information as clearly as possible and assure it supports compliance with the requirements in 10 CFR Part 40, Appendix A, Criterion 5B(6). Applicants are encouraged to follow the numbering system and headings of the standard format and use appendices for including supporting data not specifically included in a particular section. Conventional abbreviations should be used consistently throughout the application. Any abbreviations, symbols, or special terms should be defined where they first appear in the text. Where appropriate, calculated error bands or estimated uncertainties should be included along with numerical values. Some types of information are better presented clearly and concisely in a graphical manner by using maps, graphs, drawings, or tables and appropriate citations in the text descriptions. Applicants should ensure that graphical materials are legible and that the physical scales are adequate to clearly show details and notations. Symbols should be clearly defined and referenced.

Table 1 shows the standardized outline for an Alternate Concentration Limit application. It includes chapters for supporting information on the site and its setting, a hazard assessment, an exposure assessment, a review of realistic corrective action alternatives, and the proposed concentration limits.

The application should also be structured to allow ready substitution of pages in response to reviewer comments and information requests. Pages should be punched for a standard loose-leaf binder. Revisions should be provided on pages that will replace the original pages, with the changes indicated by a "line change" demarcation in the margin. The date and revision number should be indicated in the bottom outside margin of each revised page, and the package of submitted revisions should include a list of all page replacements for the application. The font style and text size should be plain and large enough to allow the document to be scanned electronically for easy inclusion in the Agencywide Documents Access and Management System (ADAMS). All figures and diagrams should also be clearly presented to assist in electronic scanning.

A legible base map is essential for all applications. The base map should include the tailings disposal area, the location of the reclaimed slopes, the Point of Compliance locations, the Point of Exposure locations, monitoring wells locations, and the proposed long-term control boundary. Pertinent site data, such as potentiometric surfaces, isoconcentration contours, and forecasted concentrations should use the base map as the common reference.

Table K–1. Standard Format of an Alternate Concentration Limit Application

EXECUTIVE SUMMARY

TABLE OF CONTENTS

NRC FORM 335
(2-89)
NRCM 1102,
3201, 3202

U.S. NUCLEAR REGULATORY COMMISSION

BIBLIOGRAPHIC DATA SHEET

(See Instructions on the reverse)

1. REPORT NUMBER
(Assigned by NRC, Add Vol., Supp., Rev., and Addendum Numbers, If any.)

NUREG-1620, Rev. 1

2. TITLE AND SUBTITLE

Standard Review Plan for the Review of a Reclamation Plan for Mill Tailings Sites Under Title II of the Uranium Mill Tailings Radiation Control Act of 1978
FINAL REPORT

3.	DATE REPORT PUBLISHED	
	MONTH	YEAR
	June	2003

4. FIN OR GRANT NUMBER

5. AUTHOR(S)

John Lusher

6. TYPE OF REPORT

Technical

7. PERIOD COVERED *(Inclusive Dates)*

8. PERFORMING ORGANIZATION - NAME AND ADDRESS *(If NRC, provide Division, Office or Region, U.S. Nuclear Regulatory Commission, and mailing address; if contractor, provide name and mailing address.)*

Division of Fuel Cycle Safety and Safeguards
Office of Nuclear Material Safety and Safeguards
U.S. Nuclear Regulatory Commission
Washington, DC 20555-0011

9. SPONSORING ORGANIZATION - NAME AND ADDRESS *(If NRC, type "Same as above"; if contractor, provide NRC Division, Office or Region, U.S. Nuclear Regulatory Commission, and mailing address.)*

Same as 8. above

10. SUPPLEMENTARY NOTES

Supersedes

11. ABSTRACT *(200 words or less)*

A U.S. Nuclear Regulatory Commission source and byproduct materials license is required by 10CFR Part 40 for the operation of uranium mills and the disposal of "tailings," wastes produced by the extraction or concentration of source material from ores processed primarily for their source material content. Appendix A to Part 40 establishes technical and other criteria relating to siting, operation, decontamination, decommissioning, and reclamation of mills and of tailings at mill sites. The licensee's site reclamation plan documents how the proposed activities demonstrate compliance with the criteria in Appendix A to Part 40 and the information needed to prepare the environmental assessment on the effects of the proposed reclamation activities on the health and safety of the public and on the environment.

This standard review plan is prepared for the guidance of staff reviewers in the Office of Nuclear Material Safety and Safeguards in performing safety and environmental reviews of reclamation plans for uranium mill tailings sites covered by Title II of the Uranium Mill Tailings Radiation Control Act of 1978 as amended. It provides guidance for new reclamation plans, license renewals, and license amendments. The principal purpose of this standard review plan is to ensure the quality and uniformity of staff reviews and to present a well-defined base from which to evaluate changes in the scope and requirements of a review.

This standard review plan is written to cover a variety of site conditions and reclamation plans. Each section contains a description of the areas of review, review procedures, acceptance criteria, and evaluation findings. Revision 1 also incorporates information to address new Commission policy on several issues related to uranium recovery.

12. KEY WORDS/DESCRIPTORS *(List words or phrases that will assist researchers in locating the report.)*

Title II sites; tailings; reclamation plan; closure activity; license amendments; license renewals; financial surety; license termination; long-term surveillance

13. AVAILABILITY STATEMENT

unlimited

14. SECURITY CLASSIFICATION

(This Page)
unclassified

(This Report)
unclassified

15. NUMBER OF PAGES

16. PRICE

This form was electronically produced by Elite Federal Forms, Inc.

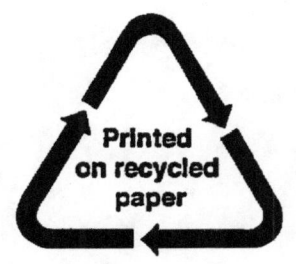

www.ingramcontent.com/pod-product-compliance
Lightning Source LLC
Chambersburg PA
CBHW080239180526
45167CB00006B/2335

* 9 7 8 1 5 0 0 1 1 3 3 8 4 *